LIVE EARTH

THE CONCERTS FOR A CLIMATE IN CRISIS ™

The Global Warming Survival Handbook

77 Essential Skills to Stop Climate Change— or Live Through It

LIVE EARTH
THE CONCERTS FOR A CLIMATE IN CRISIS ™

The Global Warming Survival Handbook

77 Essential Skills to Stop Climate Change— or Live Through It

David de Rothschild

FOREWORD BY Rob Reiner
AFTERWORD BY Kevin Wall

ILLUSTRATIONS BY William van Roden
PRODUCED BY Melcher Media

WITH CONTRIBUTIONS BY Adam M. Bright, Max Dickstein, Mikki Halpin, Kira Gould, Shelley Lewis, Aaron Naparstek, Erik Ness, Andrew Postman, Hillary Rosner, Kate Sekules, Amanda Park Taylor, Jocko Weyland, Lizz Winstead, and Michele Zipp

RODALE

MELCHER MEDIA

The authors and editors have made extensive efforts to research and ensure
the accuracy of the information contained in this book. However, the contents
are for informational purposes only, and neither the creators nor the publisher
can assume any responsibility for any accident, injuries, losses, or other damages resulting from the use of this book.

Any errors brought to the publisher's attention will be corrected in future
editions. Telephone numbers and URLs given in this book were accurate at
the time it went to press.

Rodale books may be purchased for business or promotional use or for special
sales. For information please write to:
Special Markets Department, Rodale Inc.
733 Third Avenue, New York, NY 10017

Printed in the United Kingdom by Butler and Tanner, Ltd.

Library of Congress Cataloging-in-Publication Data is on file with the publisher.

10-digit ISBN: 1-59486-781-X
13-digit ISBN: 978-1-59486-781-1

Distributed to the book trade by Holtzbrinck Publishers

2 4 6 8 10 9 7 5 3 1 softcover

RODALE

Rodale Inc. makes every effort to use acid-free ⊗, recycled paper ✪. We
inspire and enable people to improve their lives and the world around them. For
more of our products visit rodalestore.com or call 800-848-4735.

Contents

Foreword

Rob Reiner
DIRECTOR, ENVIRONMENTALIST

At Live Earth, we are sending a message to the world: S.O.S.

It's a distress call to Save Our Selves from our climate crisis. It is the most urgent message we can deliver to our friends, to our leaders, to the rest of the world.

Our climate crisis affects everyone, everywhere. The Earth's temperature is rising. Our water and food sources are becoming scarce, our oceans polluted, our coastal cities and island countries endangered. And its magnitude means that only a truly global response can address it.

The good news is that each of us can take action to solve this crisis. All of us have a role to play, and none of us bears the burden alone. We have the know-how and the technology. Cleaner cars, wind farms, solar power, even better lightbulbs.

This book outlines 77 skills we can practice today. As you will see, many of them are simple, easy, and inexpensive to perform. Some require a larger commitment of time and resources. We are all at different stages of awareness and action, and whether you are just beginning or you are an expert on this issue, this book has something for you.

Live Earth is the start of a global environmental movement, one that harnesses the power of everyone working together. So let us not be overwhelmed by the size of the problem. The positive sum of small actions, multiplied by millions of people, can lead to dramatic effects. You are part of this movement and the small changes you make will add up.

Preface

David de Rothschild
FOUNDER, ADVENTURE ECOLOGY

I have always been more interested in what's going on outside my window than inside.

This natural curiosity has been the source for some extraordinary adventures. I have had the incredible opportunity to spend many months in some of the world's most fragile ecosystems. It was during a polar adventure that I truly began to grasp the scale and complexity of climate change. Standing in the midst of the Arctic, surrounded by 14 million square kilometers of frozen ocean, I felt like nothing more than a speck of dust on the endless horizon of Earth's most raw, majestic, and environmentally significant ecosystem.

That expedition, to cross the Arctic Ocean, was halted when warm temperatures made the ice too vulnerable. And it suddenly became clear that thinking of protecting the planet was no longer enough; what we have to think about is our ability to survive on this planet. We need to have the necessary skills in order to do so.

In all of my travels, I've found that luck favors the well prepared. Hence, *The Global Warming Survival Handbook*. I hope it will prepare you with the skills needed for stopping climate change—and for living through it, should we not take the necessary steps!

It's our responsibility as "globalsapiens" to create a more sustainable path than the one we have followed. The time has come to stop being part of the problem and to become part of the solution.

Despair does not inspire action. The 77 skills described here will encourage action as well as thought. Most important, this handbook should motivate you to join the millions of others on the adventure of transforming Earth's "hot spots" into "hope spots."

Climate Control:
An Introduction to
Global Warming

Discussing the weather used to be so casual.
Nice day, huh? Or, *You believe all this rain?*

It's harder these days—even our small talk is controversial. *Scorcher, isn't it? Must be that global warming.* For many, the greatest environmental threat that humans have ever faced is a political issue.

But it's not. Global warming is a matter of scientific fact: 2006 was the sixth-warmest year on record since 1850; the top five spots are held by 1998, 2005, 2003, 2002, and 2004. Whether you're a Kiribati fisherman losing your island home to the rising sea or a snowboarder who couldn't find any snow last January, climate change is making its presence known.

If we fail to recognize the immediate nature of this threat, the consequences could be catastrophic. Rising seas, searing temperatures, killer storms, drought, plague, pestilence. Finding solutions begins with identifying the problem. Briefly, here's how global warming works. Energy from the Sun, in the form of light and heat, warms the Earth. Heat rises, and some of it heads back into space. Most of it, though, is trapped by molecules in the atmosphere—molecules of "greenhouse gases," named because their effect is just like that of a greenhouse. Water vapor is a primary greenhouse gas, in addition to carbon dioxide (CO_2), methane, and nitrous oxide. All are natural—indeed, without the greenhouse effect, the Earth would be cold and uninhabitable.

The problem is, we have greatly increased the amount of CO_2, methane, and nitrous oxide in the atmosphere—mostly by burning things, like forests and fossil fuels. Before the Industrial Revolution, there were 280 parts per million (ppm) of CO_2 in the atmosphere. Today, CO_2 is about 380 ppm. Factor in all the other man-made emissions, and the result is equivalent to 430 ppm of CO_2.

The more greenhouse gas molecules, the more heat the Earth keeps. The atmosphere is so huge, change seems to occur slowly, if at all, but these increases are adding up. Over the last 30 years, the Earth's average temperature has warmed by a full 1°F. One degree may not sound like much—but the Little Ice Age in the middle of the last millennium was signaled by a shift of only 2 to 4°F. Inertia in the system will warm us at least another 1°F by 2020.

But humans are not sitting still; greenhouse emissions are accelerating, and at current growth rates we will double the preindustrial levels of gases by 2050. If countries such as China and India adopt the same carbon crutch the U.S. relies upon, the doubling could arrive as early as 2035.

What happens then? With a doubling of greenhouse gases, a 4°F rise in global average temperature becomes likely within this century. And all kinds of nastiness kick in at that point. By 2080, up to three billion people could suffer water shortages, and 200 to 600 million could face famine. Twenty to 30% of all species may face extinction. The melting of the Greenland Ice Sheet would accelerate, raising sea levels by as much as 23 feet.

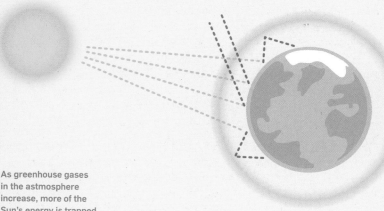

As greenhouse gases in the astmosphere increase, more of the Sun's energy is trapped.

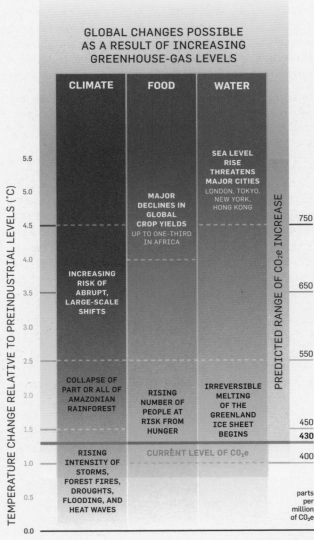

GLOBAL CHANGES POSSIBLE
AS A RESULT OF INCREASING
GREENHOUSE-GAS LEVELS

CLIMATE	FOOD	WATER

TEMPERATURE CHANGE RELATIVE TO PREINDUSTRIAL LEVELS (°C)

PREDICTED RANGE OF CO_2e INCREASE

5.5 — SEA LEVEL RISE THREATENS MAJOR CITIES LONDON, TOKYO, NEW YORK, HONG KONG

5.0 — MAJOR DECLINES IN GLOBAL CROP YIELDS UP TO ONE-THIRD IN AFRICA

4.5 — 750

4.0 —

3.5 — INCREASING RISK OF ABRUPT, LARGE-SCALE SHIFTS — 650

3.0 —

2.5 — 550

2.0 — COLLAPSE OF PART OR ALL OF AMAZONIAN RAINFOREST — RISING NUMBER OF PEOPLE AT RISK FROM HUNGER — IRREVERSIBLE MELTING OF THE GREENLAND ICE SHEET BEGINS

1.5 — 450 / **430**

CURRENT LEVEL OF CO_2e

1.0 — RISING INTENSITY OF STORMS, FOREST FIRES, DROUGHTS, FLOODING, AND HEAT WAVES — 400

0.5 — parts per million of CO_2e

0.0 —

CHART BASED ON THE STERN REVIEW ON THE ECONOMICS OF CLIMATE CHANGE

We are speeding into a troubling void. Nobody knows exactly how it's going to unfold. But the broad outlines are quite clear, absolutely devastating, and backed by evidence that almost every reputable scientist now calls *overwhelming* and *unequivocal*. These are not common words in the scientific vocabulary. But after decades of study and debate, the leading scientific organizations around the world now call climate change a real and pressing threat.

Science has given us unprecedented foresight. Now we must find the conviction to act. James Hansen, a NASA climate researcher, believes that "we have a very brief window of opportunity to deal with climate change . . . no longer than a decade, at the most." How we face the next few years could profoundly affect the rest of history. If we don't begin to change today, it may be too late.

The task ahead is monumental. It's far bigger than what one person can do—bigger even than what one country can do. Just to keep greenhouse gases at their present level will require slashing our emissions by 60%. To *lower* the level of gases and to reverse the warming already underway, the entire fossil-fuel economy that brought us to unprecedented prosperity and technological prowess will need rethinking. In the short term, we need to tweak it—throw fingers in the dike to stanch the flow of carbon. In the long term we may need to remodel, reinvent, and renew some major building blocks of society.

Your help is needed. You've got to take your share of small steps—around the house, around the office, in your neighborhood—because little things multiplied by the million add up, just as surely as individual CO_2 molecules add up. We also need you to educate your friends about policy initiatives, to support and vote for leaders who have the vision to make a difference, to help remodel our cities, and to invent the technologies of the future. The journey begins with some small talk about the weather.

The graph on the opposite page illustrates the types of impacts that could be experienced as the world heats up because of greenhouse gases. Different levels of CO_2 in the atmosphere bring a variety of large-scale effects.

10 Easy Steps to Help Fight Global Warming

Adapted from Adventure Ecology's *Top 10 for the Planet*

1 Adjust Your Climate by Two Degrees

SEE SKILL #20, PUT ON A SWEATER

Turning your heat down by 2°F in the winter and your air-conditioner up by 2°F in the summer can save our planet from more than one-third of a ton of CO_2 emissions per year.

2 Change a Lightbulb

SEE SKILL #3, REPLACE A LIGHTBULB

If every household in the United Kingdom used just one energy-saving lightbulb, it would be enough to shut down a power station.

3 Stop Appliances from Standing By

SEE SKILL #10, KILL YOUR PHANTOMS

Those little red lights on your TV, stereo, and computer? They're on standby, waiting for your remote click, and still sapping energy. Turn them off for real by unplugging them, and cut your home energy–related emissions by 10% or more.

4 Say No to Plastic Bags

SEE SKILL #38, CHOOSE THE RIGHT BAG

The story of the 500 billion to 1 trillion plastic bags used each year doesn't have a happy ending. Most end up in landfills, blowing through the streets, or hurting animals on land and in the sea.

5 Shop Locally

SEE SKILLS #29, COUNT YOUR FOOD MILES, AND #30, BEFRIEND YOUR FARMER

On average, each item in your local supermarket has traveled at least 1,000 miles to get to you. Buying locally produced food reduces the amount of energy used for transportation.

6 Bring Your Own Mug

SEE SKILLS #18, SAY NO TO STYROFOAM, AND #25, GREEN YOUR CUBE

Bring your own travel mug. Disposable cups (and lids!) go directly into landfills, where they've become a major presence.

7 Go Public

SEE SKILL #36, DECONGEST DOWNTOWN

One bus can carry the same number of people as 50 cars. Subways and trains hold even more. For every mile you travel, public transport uses around half the fuel of a private car.

8 Bike or Walk

SEE SKILL #35, RIDE A BIKE

Bike or walk to work, to school, to the store—doing so just once a week will give the planet a much-needed break from the CO_2 emissions of your car.

9 Say Yes to Short Showers

SEE SKILL #45, TAKE A BATH TOGETHER

Global warming promises to worsen water shortages around the world. A quick shower uses one-third the water of a bath. Cutting your shower time by one minute can save more than 500 gallons of water each year.

10 Plant Something

SEE SKILLS #28, GROW YOUR OWN TOMATO, #39, PLANT A TREE (MINDFULLY), AND #48, GREEN YOUR ROOF

Plants take in CO_2 and pump out oxygen. A single tree provides enough oxygen for two people for their entire lives. Plants and trees provide homes and food for birds and other wildlife.

0 | How to Read This Book

Scale it up: if one million people changed one thing in their lives, the result would be...

COST*	TIME*	EFFORT*	IMPACT*
$20	15 min		

FOOD AND SHELTER

SURVIVAL STRATEGY

ENERGY AND TRANSPORTATION

WORK AND COMMERCE

IF ALL ELSE FAILS

Bright idea

CO_2 alert

Danger

Energy efficiency

$$$ Money-saving tip

The atmosphere is thickening. The planet is on fire. What do you do?

This book is a survival guide. The 77 skills described here offer many different ways you can help put out the fire. There are small steps to take, there are big steps, and then there are steps that will be necessary only if all else fails. Those are coded in high-alert pink stripes.

Start with the first—"Commit" to the fight—or choose one from the middle. Every step is a step for the planet. Every step is a part of our survival.

*Cost and time are minimum estimates. Your results may vary. Personal effort and potential impact are measured on a scale of 1 to 5.

DISCLAIMER: If all else fails...
We hope it never gets this bad. But if it does, Live Earth cannot guarantee that space will be habitable or that your camel will be friendly. Please take these skills not so much as literal advice but as impractical responses to unfathomable situations.

1 | Commit

	COST	TIME	EFFORT	IMPACT
If one million people stand up to say, "I can help fight global warming," we have taken the first step.	$0	24/7		

The two most critical survival skills are clear thinking and perseverance.

Global warming may be the most serious challenge the human race has ever faced, but don't freak. Your survival depends on your ability to stay calm, to use your head, and to stick it out.

If this book is your first exposure to the facts about global warming, you may slip into a state of "climate shock"— triggered by the sudden realization of the imminent dangers we face. Warning signs include shortness of breath, a sense of powerlessness and despair, a concern for polar bears, and an urge to lecture others about glaciers.

Luckily, climate shock is manageable—and so is climate change, if all of us are committed to the task. A first step: memorize this acronym.

C	ommit this book to memory
H	unt down and destroy your excessive CO_2 habits, one by one
I	nspire others to do the same
L	obby your leaders to make big changes
L	augh a little—saving the planet is going to be tough, but it doesn't have to be boring

2 | Change Slowly

	COST	TIME	EFFORT	IMPACT
If one million Americans reduced their emissions by 10%, we'd eliminate 750,000 tons of CO_2 per year.	$0	just a little		

Arguably the most celebrated aphorism about incremental change—"A journey of a thousand miles begins with a single step," alternately attributed to Confucius, Mao, or anonymous Chinese proverbial wisdom—is likely mistranslated. The statement refers to "a thousand *li*," roughly 400 miles.

Most attempts at change fail—not for lack of will, but for lack of planning and pragmatism. And yet the behavior you wish to change didn't embed itself in a day, so why should you expect to undo it in a day?

PICK ONE GOAL Review its pros and cons. To make this change, will you be forced somehow to compromise your values? One goal to consider: reducing your carbon footprint by 1,000 pounds this year.

WRITE IT. BE CLEAR ABOUT IT. IS IT REALLY ACHIEVABLE? "I want my home to be 100% reliant on solar power by next Thursday" is an idealistic wish. "I want to decrease my energy consumption of fossil fuels by 10% in six months" is the start of a project.

IDENTIFY THE SKILLS AND RESOURCES NEEDED If you haven't ridden a bike in 20 years, don't sell everything you own with four wheels and an engine and rely on people power just yet. Find alternatives to how you do things now—for example, could you carpool one or two days a week? Examine why you continue to engage in certain behaviors (convenience?). How can your new way of being accommodate your preferences and needs?

DIVVY YOUR PLAN INTO SMALL, ACHIEVABLE STEPS
("On Monday, I'll call my energy company and switch
to that plan that helps to subsidize greener energy alternatives.") Each success builds confidence. If you fail to
achieve a step, come up with a more achievable one.
(And don't compare yourself to what you think others
might achieve; it's your life you're changing.)

ANTICIPATE HICCUPS Life will inevitably intervene, so
plan for it, and for how you'll regroup when that happens.
"While my family was visiting, they insisted that we heat
the house like a toaster. But this week, I'll turn the thermostat down one degree."

TALK ABOUT IT Tell friends and family what you're up
to. Emphasize one really cool thing about it. This makes
the goal more concrete for you, plus they can
offer support.

REVIEW THE PLAN Are you making progress? Do you need to modify your plan?
Note which parts work well, tweak those
that don't.

TREAT YOURSELF As you hit little marks
of success, acknowledge the work you're
putting in to make the change. No shame
in feeling good; the better you feel, the
more you're capable of doing.

KEEP YOUR EYES ON THE PRIZE Changing an
established way of life can take three to six
months or more. Don't give up if you're still
struggling at four weeks. In time, it gets
easier and, eventually, becomes innate.

3 | Replace a Lightbulb

If one million households changed four lightbulbs each, 900,000 tons of greenhouse gases would be eliminated.

COST	TIME	EFFORT	IMPACT
$3	10 min		

Convince every American household to replace at least one of their incandescent bulbs with a CFL and prevent greenhouse gases equivalent to the annual emissions of nearly 800,000 cars.

500 lbs

is the equivalent amount of coal burned when you choose an incandescent bulb over a compact fluorescent, over the life of the bulb.

Watt!?

Lighting uses more energy globally than is generated by nuclear or hydro-power.

Early compact fluorescents got a bad rap because their ghostly, grim light made you and your friends look like the walking dead. Thanks to improved phosphors, it's now easy to find CFLs that produce a warm, natural glow.

For the most lifelike results, look for bulbs with a color temperature of about 2,700K, and a color rendition index of at least 80. Make sure the bulbs are Energy Star–approved.

IDENTIFY

Find an incandescent bulb in your house. It should be easy to do. It's the same type of lightbulb that we have been using since 1878.

SHOP

Head to the store. Spend about $2.60 on a compact fluorescent lightbulb (CFL) to replace the old bulb. (The new bulb will look like a soft-serve vanilla ice-cream cone.)

TURN IT OFF

Make sure you have the light switch turned off. (Ideally it's daytime, so you can still see.)

REMOVE

Carefully turn the old bulb counterclockwise until it releases from the light socket. Save it for emergencies.

REPLACE

Insert new compact fluorescent lightbulb. That's the one that provides the same amount of light, lasts up to 10 times longer, and uses about 75% less energy than your old bulb.

SAVE

Deposit $30 to $50 of electricity savings in the bank over the life of the bulb. This holds true for every bulb you replace, so why not switch a dozen?

The annual personal carbon footprint of one million Americans equals the CO_2 emissions of 1.2 million cars.	COST	TIME	EFFORT	IMPACT
	$0	10 min		

Some Web calculators will offer to sell you lightbulbs and carbon offsets. Exercise your customary skepticism.

1 ton is a great goal for the amount of CO_2 to shave off your annual carbon usage.

Everyone is part of the problem. The question is: how much of a problem are you? There are hundreds of websites to help measure the carbon footprint of every step of your life—even that last cheeseburger.

Most calculations focus on the three big-carbon-ticket items in our lives: cars, planes, and houses. Residential energy use accounts for 21% of U.S. carbon emissions, while transportation accounts for 33%. Air travel makes up about 10% of that, but frequent fliers' footprints get big quickly.

Every calculator makes assumptions about your life, so read their explanations. For instance, many assume you live in Europe. Try a few different models and compare the results; if they differ radically, figuring out why may tell you a lot about your personal carbon output.

Local calculators often work better—for example, one for Alaska should factor in renewable energy in different parts of the state. The best calculators, like the EPA's, tackle more complex issues such as recycling, food miles, or how much you'll accomplish by lowering your thermostat. Find that calculator at www.epa.gov/climatechange/emissions/ind_calculator.html.

A SIMPLE WORKSHEET

There are many other contributors to your carbon bottom line, but these four boxes will measure about half of your footprint. And it's not simply CO_2—these figures give you a "CO_2 equivalent," which takes methane and other greenhouse gases into account.

🚗 DRIVING

MILES DRIVEN PER YEAR				A
CAR'S AVERAGE GAS MILEAGE				B
A	÷	B	=	C
19.654 (CO_2e/gal)	×	C (gal / year)	=	D

✈ FLYING

MILES FLOWN PER YEAR				E
1.36 ($CO_2e/mile$)	×	E	=	F

⚡ ELECTRICITY

KILOWATT HOURS USED PER YEAR				G
1.37 (CO_2e/kWh)	×	G	=	H

🔥 NATURAL GAS

CCF (100 CUBIC FEET) USED PER YEAR				J
12.0593 (CO_2e/CCF)	×	J	=	K

TOTAL LBS CO_2 EQUIVALENT PER YEAR

D	+	F	+	H	+	K	=	TOTAL

RATE YOUR FOOTPRINT

FREQUENT FLIER	AVERAGE JOE	TREE HUGGER	OUR GOAL
80,000	15,000	3,300	0

Note: Formulas on this page are based on the Bonneville Environmental Foundation's carbon calculator at www.b-e-f.org.

5 | Sub-Size It

	COST	TIME	EFFORT	IMPACT
If every American lost one pound, we'd save 39 million gallons of fuel per year.	$0	0		

A wide-screen TV can use more electricity than your refrigerator.

1 in 5

new U.S. homes is larger than 3,000 square feet.

The American dream has been supersized—and we've bitten off more than we can chew. With less than 5% of the world's population, the U.S. creates nearly one-quarter of the Earth's greenhouse gases. And American-style products—from SUVs to McMansions—are spreading around the globe.

As consumers, Americans believe that size matters: given a choice, we make it extra large. It's not just French fries. From 1970 to 2004, the average new home in the U.S. grew more than 50%, from 1,500 to 2,349 square feet, even as the average number of people in each household shrunk by 17%.

Choosing big adds up. Manufacturing a big car creates more greenhouse gases—and that big car keeps creating more until it hits the junkyard. Cars and televisions last 10 to 15 years. Houses can last 100 years or more. The decisions we make now can affect greenhouse-gas pollution for decades.

Smaller choices are almost always better for the planet—and usually cost less, too. Less maintenance time, lower finance charges, easier parking, cooler planet.

SUPER-SIZED VS. SUB-SIZED

McMANSION

Per capita residential energy use has risen about 8% since 1985.

SMALL HOUSE

Houses between 1,500 and 2,000 sq ft consume 40% less energy than a 4,000+ sq ft McMansion.

HUMMER H2

6,400 lbs
11/14 mpg city/highway (est.)

SMART CAR FORTWO

1,606 lbs
58.8 mpg (UK model est.)

PLASMA

TVs account for about 4% of U.S. energy consumption—but that could double if everyone goes big-screen. A 42" plasma screen consumes up to 400 watts.

LCD

LCDs use about half the power of plasma TVs, but at 100 watts, an old-school 27" CRT uses even less.

If one million drivers stayed parked for just one day, 20,000 tons of CO_2 emissions could be eliminated.

COST	TIME	EFFORT	IMPACT
$0+	5+ min		

STARTER

Allowing your car to idle for 10 seconds uses more gasoline than turning the car off, then on again.

WEIGHT

A heavier car needs more gas. Add just 100 extra pounds and you're going to burn 2% more fuel. Unless your office is in Fallujah, don't drive to work in a Hummer. In fact, don't even sit in a Hummer.

ENGINE

A simple tune-up can improve your car's fuel efficiency from 15 to 50%.

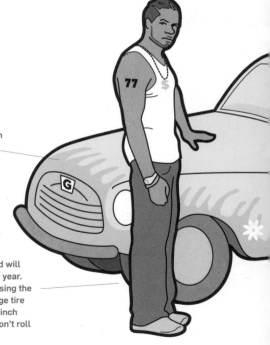

TIRES

Keeping your tires properly inflated will save you about one tank of gas per year. Check the tire pressure monthly, using the same gauge every time. The average tire loses about one pound per square inch every month. Underinflated tires don't roll as efficiently and wear out faster.

GAS AND BRAKE PEDALS

While it can be fun to pretend you're a cabdriver, avoid jackrabbit starts and jamming on the brakes. Aggressive driving can increase fuel consumption by 33%, and spews five times more toxic emissions than driving slowly and smoothly. If you keep it at 55 mph, you'll use 15% less fuel than if you go 65 mph.

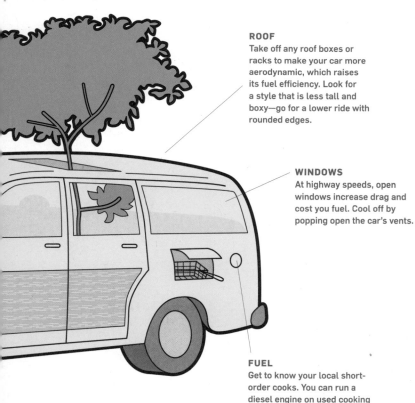

ROOF

Take off any roof boxes or racks to make your car more aerodynamic, which raises its fuel efficiency. Look for a style that is less tall and boxy—go for a lower ride with rounded edges.

WINDOWS

At highway speeds, open windows increase drag and cost you fuel. Cool off by popping open the car's vents.

FUEL

Get to know your local short-order cooks. You can run a diesel engine on used cooking oil or bio-diesel fuel. (See Skill # 56.)

7 | Fly Right

If one million passengers decided to skip one five-hour flight, one million tons of CO_2 would be eliminated.

COST	TIME	EFFORT	IMPACT
$0	none		

The most fuel-efficient flight length is 2,700 miles.

If you're looking to dump a whole lot of warming gases into the atmosphere really quickly, then flying aboard a jet airplane is a great way to do it.

A one-way flight from New York to Los Angeles aboard a typical commercial airliner makes each passenger responsible for one ton of CO_2 emissions in six short hours. By comparison, you'd have to drive a full day and a full night in a Ford Explorer to produce that much CO_2.

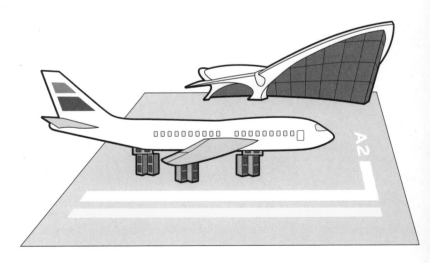

But the problem is even worse than it seems. Most of those emissions occur at high altitude, where CO_2 wreaks no greater harm than at ground level; but airplane exhaust's nitrous oxide and water vapor sure do. Contrails—those pretty, white water-vapor ribbons in the sky—trap heat in the atmosphere as efficiently as CO_2 does and may also encourage the formation of heat-trapping, high-altitude cirrus clouds.

The first line of defense: fly less. Trains, buses, and even cars are gentler on the atmosphere. Of course, it may be unrealistic to forswear jet travel permanently. What are the best options?

5,400 lbs
of CO_2 emissions are released per passenger flying from New York to Tokyo.

TIPS FOR THE NEW JET TRAVELER

Fly direct whenever possible. Landing and taking off are the most gas-guzzling parts of your flight; takeoff can consume 25% of the total fuel on short flights. If you must fly, fly long. Try to avoid flights of 600 miles or less.

Take the train instead. Seriously. (See Skill # 63.) Particularly for shorter trips, where the time difference is less pronounced, rail is the only way to go.

Fly during the day. One study shows that nighttime contrails create more warming.

Fly a newer, more fuel-efficient plane, such as the Airbus A340 or A380 or a Boeing B787. More demand will bring more efficient planes.

Offset your emissions. Just can't quit flying? Buy carbon offsets. (See Skill # 8.)

8 | Buy Carbon Offsets*

*Or not. All carbon offsets are not created equal—read on.

COST	TIME	EFFORT	IMPACT
$5+/ton CO_2	10 min		

ROAD TRIP TO THE HOOVER DAM
2,000 CREDITS

HAPPY HOLIDAYS

HOLIDAY DISPLAY
99,000 CREDITS

Unless you live in the woods and never drive or take a plane, you will need some help getting your carbon footprint down to zero.

Let's say that despite your best efforts, you still have to fly to your best friend's wedding. You're dumping 3,000 pounds of CO_2 into the atmosphere, and you're wracked with guilt about your contribution to global warming. Relax, you can throw money at the problem. Go online, find a company that sells clean energy credits, and buy enough to make up for the greenhouse gases your trip created. Problem solved? Are carbon offsets the right place to turn? The short answer is, Yes. Maybe. If you choose your offsets carefully.

A key issue is whether offsets demonstrate "additionality." That is, would a particular clean-energy project or emissions-reducing initiative have occurred without your financial contribution? If so, it's really not an offset.

In the wild frontier of carbon offsets, additionality is difficult to guarantee or even to discern. To make matters worse, there is a glut of offset competition. A few places to start: the nonprofit CarbonFund; NativeEnergy, which funds renewable-energy projects that help communities in need; and the UK's CarbonNeutral Company, which has a stringent written protocol for the use of offsets.

Many believe that carbon offsets are like a sin tax—you pay the extra money, you keep driving your SUV. The critics say that's abusing the system. Bottom line: carbon offsets should be used to compensate for that last little bit of greenhouse-gas emissions you can't eliminate any other way. They're a supplement to the effort, not a substitute. Here are some common offset options.

ROUND-TRIP FLIGHT TO VEGAS
3,000 CREDITS

TYPES OF CARBON OFFSETS

FUTURE CLEAN ENERGY PROJECTS

✓✓ Your money gets channeled toward the building of new projects that will create clean energy—wind farms, solar plants, biomass facilities that capture methane, and so on.

SUSTAINABLE DEVELOPMENT PROJECTS

✓✓ Your money supports energy-efficient and eco-friendly construction, typically in the developing world where such investments are more cost-effective.

EMISSIONS-REDUCING INITIATIVES

✓✓ Your money helps launch clean-energy education programs and lobbying initiatives.

CREDITS IN EXISTING CLEAN ENERGY

✓ Your money buys Renewable Energy Credits (RECs), also known as "green tags," each of which represent the delivery of one megawatt-hour of renewable power to the grid. There is no standard way to verify that your RECs are additional.

BUNDLED PORTFOLIO OF CLEAN-ENERGY PROJECTS

✓ Your money is invested in a selection of high-quality projects in order to hedge against the possibility that one planned project or another may never get off the ground.

REFORESTATION PROJECTS

✗ Your money pays for replanting forests or creating new ones in order to sequester carbon. Yet organizations such as Greenpeace and Friends of the Earth have criticized the practice, arguing that newly planted trees are at best a short-term carbon "loan" that doesn't reduce total carbon emissions because trees die and release what they've stored. (See Skill # 39.) Nevertheless, reforestation is complex, and evaluations of its usefulness continue to evolve.

9 | Imagine

In the future, millions of lives will be changed by global warming. How they change is up to you—and everyone around you.

COST	TIME	EFFORT	IMPACT
$0	1 hr		

We know that rising temperatures threaten to remake the world. But how can we predict the shape of that warmer world 50 years from now? We can't even forecast if it will rain next week.

One approach to seeing the future is through scenarios—carefully crafted "what if?" stories that let us imagine several different outcomes. Shell Oil made scenarios famous when its futurists predicted, in general outlines, the early-1970s energy crisis. The goal of scenario planning is not to correctly predict the future—that's a fool's game—but to exercise the imagination so that you're ready when reality strikes.

The U.N.'s Millennium Ecosystem Assessment asked more than 1,000 economists and scientists to examine how natural systems are responding to challenges like climate change. More than 100 of these experts focused on future scenarios, imagining different paths we might take, and how the planet might respond.

You can read plenty about scenario planning, but there is no substitute for actually doing it. Why not pool the imaginations and experiences of your friends? Re-create the U.N.'s planning at a "scenario party" and imagine your own futures, or rework these four for your region. If you and your friends have absolutely nothing new to talk about and you have exhausted the complete charades repertoire, this is the perfect way to cut loose.

FOUR VISIONS OF THE FUTURE

GLOBAL ORCHESTRATION Large-scale economics drive this future, with prosperity creating solutions to environmental problems: "Supra-national institutions are well placed to deal with global environmental problems, such as climate change and fisheries," but their reactive approach "makes them vulnerable to surprises."

ORDER FROM STRENGTH It will be a fragmented world, focused on security and protection. Here, people "see the environment as secondary to their other challenges." This future may promise security, but the experts found it "unsustainable and ultimately disastrous."

ADAPTING MOSAIC Politics and economics play out on a local and regional level—and some regions do a better job than others. At first, people ignore global challenges, but local successes gradually grow into larger solutions. Still, "global environmental surprises become common."

TECHNOGARDEN A world connected by environmental technology, both small and very large scale. The TechnoGarden is "a globally connected world relying strongly on technology and on highly managed and often-engineered ecosystems."

10 | Kill Your Phantoms

	COST	TIME	EFFORT	IMPACT
If one million households halved their phantom power load, we'd eliminate 150,000 tons of CO_2 per year.	$0	2 min		

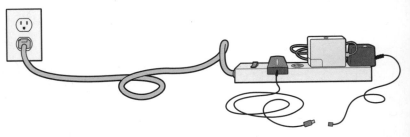

$$$

Buy Energy Star appliances when you can. (See Skill #58.) You'll cut the phantom-energy load by up to 50%. You can find the label on major appliances, electronics, office equipment, lighting, and even buildings.

Pull the plug

It's estimated that only 5% of the power drawn by cell phone chargers is actually used to charge phones, so the other 95% is wasted when they are left plugged in.

You may believe that when you turn an appliance off, it is, in fact, off. But most of your household appliances feature a clock, or a digital timer, or a remote control, or a standby mode....

When they're not on, they're still using electricity. It's known as phantom electricity or, more aptly, vampire power, because of the soul-sucking waste of energy and cash it causes. The phantom electrical load in the industrialized countries alone accounts for 75 million tons of CO_2 emitted and billions of dollars wasted per year.

OKAY, SO WHAT DO YOU DO? Plug as many of your electronics as possible into a power strip and then turn it off when they are not in use. TVs and VCRs alone waste $1 billion in lost electricity in the U.S. annually.

11 | Bank Online

	COST	TIME	EFFORT	IMPACT
If every U.S. home received and paid its bills online, annual greenhouse-gas emissions would drop by 2.1 million tons.	$0	5 min		

Some of us find paying bills soothing. The signing of the check, the scribbling of the return address, the choosing of just the right stamp.

It's time to snap out of it, pen pal, and enter the modern world of online bill paying. Look at your bills this month. More than likely, every single one has a paragraph begging you to pay your bill online. Online transactions save companies money they'd otherwise be spending on paper, printing, and postage.

That's their angle. Yours is saving the world for future generations. The cost in fuel and CO_2 for one bill is tiny, but it adds up. By one estimate, if each U.S. household viewed and paid its bills online, we would save more than 29 trillion BTUs of energy and reduce 1.7 billion pounds of waste a year.

You can arrange individual payments at the companies' websites, or centralize the process by setting up online banking at your financial institution. Your bank may charge you for the service, but most utilities do not.

Sign up for direct deposit if your company offers it. You'll save that emissions-producing trip to the bank.

405,000

trees are felled annually to make the paper required to send everyone in America their phone bills.

12 | Throw a Party

Twenty million Americans celebrated the first Earth Day in 1970; 200 million people worldwide marked the day in 1990.

COST	TIME	EFFORT	IMPACT
$0+	2 hrs		

There are actually two Earth Days. One is the familiar April 22 celebration. The other, celebrated by the United Nations, occurs every March and honors the spring equinox.

Sometimes the best way to raise consciousness is by raising a glass—and what deserves a toast more than our venerable old planet?

EARTH DAY The first Earth Day, held on April 22, 1970, has been called the birth of environmentalism in America. Traditional activities range from tree plantings and river cleanups to huge rallies.

CLEAN UP THE WORLD DAY The third weekend in September, join 35 million people in 120 countries on a planet-wide beautification campaign. Sign up at www.cleanuptheworld.org.

WORLD CARFREE DAY On September 22, spend a day without driving. Get a temporary street closing from your local traffic department and have a picnic on the hood of your ride. (See www.worldcarfree.net.)

13 | Get Hitched

	COST	TIME	EFFORT	IMPACT
If one million couples halved the CO_2 cost of their weddings, we'd eliminate 7.25 million tons of the stuff.	varies	'til death…		

In addition to drunken toasts, embarrassing relatives, and your ex, now there's one more thing to try to keep out of your wedding—carbon emissions.

Don't stress if you can't zero out your wedding's CO_2—your goal is to make a magical day even more meaningful, not to save the world with one nuptial.

RINGS The mining biz is an energy suck. Work around the problem by choosing vintage rings or a contemporary one using recycled gold. (Check out www.greenkarat.com.)

RSVP Drop the reply cards and reduce about a quarter of the paper you're using. Replies by email, by phone, or on a custom website are easier to track anyway.

LOCATION Save the emissions. Choose a locale that is easy to get to for the majority of your guests. (Wedding car-sharing can also lead to love connections for your unattached guests.)

Give your guests wildflower seeds to throw, not rice. Choose flowers from a local grower. For the favors, who could complain about compact fluorescent lightbulbs?

NO PUBLIC DUMPING $500.00 FINE

14 | Green Your Home

	COST	TIME	EFFORT	IMPACT
If one million people upgraded to an Energy Star fridge, we'd eliminate 556,000 tons of CO_2 emissions per year.	$20+	varies		

The average home in the U.S. creates twice the CO_2 of a car over the course of a year—about 22,000 pounds. A few retrofits will get your home, and the planet, breathing easier—and save you some serious cash. A superefficient house can cut your bills—and emissions—by 66%.

Make it automatic

A programmable thermostat will keep your home at the correct temperature all day, all year. Every 2°F you lower your thermostat in the winter can save 4% on bills—and reduce emissions by the same amount. (Don't forget your sweater.)

WINDOWS
One of the most effective household retrofits is new windows. Double-paned models with a "low-e" coating that blocks heat and UV rays can drop energy use by 20 to 30%. A good weatherproof seal around your windows is essential.

INSULATION
Your insulation can probably be improved. Look for high "R-value" options—the higher the R, the better the resistance to heat flow. Eco-friendly options include blown-in cellulose, recycled denim, and foams like Icynene.

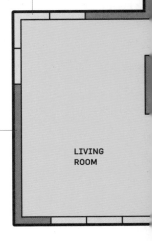

LIVING
ROOM

YARD
Well-placed plants and trees provide shade and block winds—cutting A/C and heating bills by 10 to 15%. Plus, each tree will absorb one ton of CO_2 over its lifetime.

APPLIANCES

The EPA and the European Union make this a no-brainer: look for the Energy Star label on electronics, refrigerators, washers, and air-conditioners. Penny-wise, pound-foolish: the money you'd save on cheaper appliances, you'll eventually spend on energy bills.

WATER HEATER

Wrapping your old water heater in a cozy blanket can reduce your emissions by 1,000 pounds of CO_2 a year. Blanket insulation kits can be had for less than $20.

BATHROOM

If you really want to prove your green cred, be the first on your block to install a composting toilet, such as Sancor's Envirolet.

ITCHEN

FLOORS

Consider certified woods, natural linoleum, cork, bamboo, or an earthen floor. Each has environmental benefits—from the all-natural ingredients of linoleum to the materials preserved by opting for a packed-dirt floor.

BEDROOM

LAUNDRY

Save up to 700 pounds of CO_2 by air-drying clothes for six months of the year. Double that if you're in a no-snow zone.

15 | Do the Worm

	COST	TIME	EFFORT	IMPACT
If one million people vermicomposted for a year, 82,000 tons of food waste could be turned into fertilizer.	$19/lb	5 min/wk		

The organic process of composting does produce CO_2—but that's instead of the methane the same waste would create decomposing at the landfill, which is 23 times worse for the atmosphere.

Why compost?

Your garden will love you. Compost improves plant and root growth, helps the soil hold moisture, binds nutrients in the soil, and can help inhibit plant diseases.

Do not underestimate the lowly worm; it is one of nature's best recyclers. Give hungry worms access to your kitchen vegetable waste and soon, through the process of vermicomposting, your garbage will be transformed into nutrient-rich, nonpolluting fertilizer for your thriving organic garden.

By composting with *Eisenia fetida*—worms—you'll reduce the waste you contribute to landfills, help keep methane out of the atmosphere, convert your scraps into an environmentally desirable product, and give some worms a home.

In the process of consuming waste, such as eggshells, coffee grounds, fruit and vegetable scraps, and tea leaves, worms add micronutrients, magnesium, and calcium carbonate to the organic matter that passes through their gizzards and intestines. The end result is what's known variously as vermicast, worm humus, or worm manure. This vermicast is great fertilizer for your vegetable garden or your houseplants.

VERMICOMPOSTING IN FIVE EASY STEPS

STEP 1: MAKE A WORM BIN You'll need about one square foot of surface area for each pound of vegetable waste your household produces per week. The simplest bin is an 8-to-12-inch-deep plastic or wooden box with a hinged lid and holes in the bottom for drainage.

STEP 2: ADD BEDDING Use shredded newspaper or buy commercial bedding made from coconut fiber.

STEP 3: ADD WORMS You'll need one pound of worms per square foot of surface area. (Get your worms on the Web, or check for local sources.)

STEP 4: ADD GARBAGE Worms love all vegetable matter, cooked or raw—just bury the food in the bedding. Feed your worms daily or weekly—they really don't keep track of time very well. (Don't compost meat, bones, fat, or dairy.)

STEP 5: HARVEST YOUR HUMUS Every three to six months, simply move the castings to one side of the bin and add fresh bedding. Mix the castings—gardeners call them "black gold"—with potting soil to make a super-soil, or just use it straight as fertilizer. Enjoy the cycle of life!

Pound for pound

Since worms can consume their own weight daily, six pounds of worms will make their way through six pounds of vegetable waste in one day. But be warned: six pounds equals about 6,000 worms.

	COST	TIME	EFFORT	IMPACT
One thousand 1.5 MW wind turbines would eliminate 2.7 million tons of CO_2 per year.	$.03+/ kWh	15 min		

Ask first: your utility may offer only some of these choices. No green power available? Make sure they know you want it.

Your PlayStation doesn't care whether it's juiced by Wyoming coal or a solar panel on the outskirts of Vegas. But you are a beautiful, discerning, and independent-minded energy consumer, and you deserve to connect with a power partner that brings out the best in you.

In the U.S., more than 600 utilities offer the opportunity to pay a little extra to purchase green power such as wind, biomass, and landfill gas. Similar choices are available in Canada, Australia, and much of Europe. Check your power bill or utility's website for details.

PHOTOVOLTAIC (AKA SOLAR)

WHAT IT IS	Converts sunlight to electricity
POTENTIAL	Costs drop as efficiency rises; could make 16% of world electricity by 2040
LIMITS	Expense; Sun works part-time
CO_2 CHECK	Carbon-free in 3 to 4 years

SOLAR THERMAL ELECTRICITY

WHAT IT IS	Mirrors focus the heat of the sun to power a generator
POTENTIAL	Harvesting the sun hitting 1% of the Sahara Desert could meet global electricity needs
LIMITS	Tropical is better
CO_2 CHECK	Carbon-free in less than a year

WIND

WHAT IT IS	Really big windmills
POTENTIAL	20% of available wind could satisfy global energy demands
LIMITS	Landscape impact; not all sites suitable
CO_2 CHECK	Carbon-free in less than a year

GEOTHERMAL

WHAT IT IS	Tapping the earth's internal heat
POTENTIAL	25% of Hawaiian electricity is geothermal
LIMITS	Not all sites suitable
CO_2 CHECK	Best energy return on investment

HYDROELECTRIC

WHAT IT IS	Flowing water spins a turbine
POTENTIAL	In the U.S., hydro capacity could increase by 50%
LIMITS	Most prime sites are already taken
CO_2 CHECK	Well-designed plants produce near-zero emissions

BIOMASS

WHAT IT IS	Gas from plant-derived waste and landfills fuels generators
POTENTIAL	Biorefineries could also produce fuels and chemicals
LIMITS	May require planting crops for energy
CO_2 CHECK	Close to neutral: the CO_2 emitted is absorbed from the air

COAL WITH SEQUESTRATION

WHAT IT IS	CO_2 emissions sequestered in the earth or oceans
POTENTIAL	Could capture 80 to 90% of a coal plant's CO_2 emissions
LIMITS	CO_2 storage areas are finite
CO_2 CHECK	Could be 15 to 55% of reductions needed for atmospheric stabilization through 2100

NATURAL GAS

WHAT IT IS	Fossil fuel powers generator
POTENTIAL	Cleanest fossil fuel available
LIMITS	Nonrenewable; finite
CO_2 CHECK	Produces half the CO_2 of a typical coal-fired plant

17 | Talk to Your Kids

Once upon a time, one million parents inspired one million children to look at the environment in a fresh, hopeful way.

COST	TIME	EFFORT	IMPACT
$0	20 min		

You're shocked to find yourself wearing shorts and flip-flops in February, to see footage of ice blocks the size of Delaware break off of Antarctic land masses, to visualize your hometown becoming the next lost city of Atlantis.

Problem is, kids learn from your behavior, including when and how much to freak out. If you're constantly fretting, how can you keep them positive about their future?

Don't start outfitting a bunker or having your kids fitted for hazmat suits just yet. Here are a few good tips to help them deal with their fears about global warming.

HOW TO GIVE YOUNG PEOPLE THE LOWDOWN ON CLIMATE CHANGE

FILTER Provide your child with what psychologists call "calm, unequivocal, but limited information." Filter how information comes in, at least at home. Alarming images and rhetoric are inescapable, but they should not bombard kids daily.

ENCOURAGE YOUR CHILD TO SHARE HIS OR HER FEARS Listening is helpful—kids need your attention as well as your support. Older kids will ask more questions and display more cynicism, but they still benefit from having your ear.

PUT THE NEWS IN CONTEXT With so many commentators speaking alarmingly or dismissively about climate change, it's important for you to discuss how multiple, often contradicting views can coexist.

BE MODERATE, NOT OBSESSIVE Dire as the climate crisis may be, it's moving slower than the perils of pre-adulthood—bullying, homework, swift-footed monsters.

BE POSITIVE Highlight all that they *don't* have to worry about.

FIND WAYS YOUR CHILD CAN HELP Donating time, effort, and ideas will give children (and parents) a sense of control, especially if it's something tangible (e.g., planting a tree). You're preparing children for a lifetime of approaching problems constructively.

GET ACTIVE Play catch, fly a kite, ride a bike. Get their minds off the apocalypse and let the endorphins flow. Hey, they're kids.

18 | Say No to Styrofoam

If one million people reduced their trash by 10%, we'd keep 600,000 tons of CO_2 out of the atmosphere per year.

COST	TIME	EFFORT	IMPACT
$0	1 min		

25 billion

Styrofoam cups are thrown away by Americans each year.

Sure, it keeps your coffee hot, but a thousand years after your last gulp, your empty cup will still be around.

It shouldn't take much convincing to Say No to Styrofoam, but there's so much of it! We are almost as addicted to Styrofoam as we are to oil. (Indeed, it takes 3.2 grams of fossil fuel to make a single Styrofoam coffee cup.) The U.S. makes some three million tons of it each year, the majority of which goes to the landfill.

The challenge isn't just coffee cups and packing peanuts: we love overpackaging in all shapes and materials. Whether it's for a toy, a DVD set, or a blister-packed memory card, we love big boxes. It's time to say no to all that waste. Reducing your packaging trash by 10% could save the atmosphere 1,200 pounds of CO_2 per year.

HOW TO FIGHT OVERPACKAGING

LEAVE THE PACKAGING AT THE STORE Unpack your product right there at the store and hand all that pesky packaging to the manager. This sends a message to retailers to downsize their waste.

MAKE YOUR OWN PACKING PEANUTS When packing your own boxes, pop real popcorn with an air popper and use that for packaging. The recipient can sprinkle it in their backyard for the birds.

RECYCLE While Styrofoam is hard to recycle, most plastics are not. Making one ton of recycled plastic—versus the virgin stuff—saves 685 gallons of oil.

KEEP SILVERWARE AND NAPKINS BY YOUR DESK Tell the takeout joint to hold the plastic forks and paper napkins wrapped in plastic.

GIVE POSITIVE FEEDBACK Praise companies that have switched to smaller, biodegradable, or recycled packaging.

Find Styrofoam recyclers at www.epspackaging.org/info.html

19 | Advertise Your Trash

	COST	TIME	EFFORT	IMPACT
One million clay bricks have embodied energy equal to 9,372 gallons of gasoline. Trade 'em in!	$0	15 min		

Because you're dealing primarily with strangers, use some common sense. Of course, 99.9% of the people you'll meet are interesting, responsible, environmentally concerned folk like yourself. If you think you may be dealing with the other .1%, just politely decline, or opt to meet in public—and make sure to get any cash up front.

One man's tacky eyesore is the pièce de résistance of another man's Hawaiian-themed den. So strike a deal: advertise your unwanted items on the Internet and sell them to someone who wants them.

You'll give the local landfill some breathing room, save or make some money (the exact amount depends only on your generosity, charm, and bartering prowess), and come away with a closer sense of community. You may even make a friend as part of the bargain. After all, you and this complete stranger already have something special in common: your fondness—whether it's waxing or waning—for that old ceramic bust of Tom Selleck.

Because it's free, the Internet is the best place to list old items, whether it's to auction off your brass thumbtack collection for a cool G on eBay, or to sell Grandma's floral ottoman ("needs a good reupholstering") for a tidy profit on Craigslist. Online markets are great for creating instant links between buyers and sellers, and they also connect like-

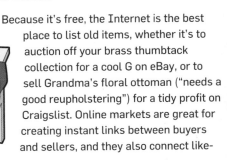

minded swappers. Craigslist, for example, has a section (labeled "free") where users list free items that anyone's welcome to pick up, and the same basic deal goes for the younger site www.freecycle.org, which also allows you to post "wanted" notices, requesting free stuff you'd like to receive. Where else can you ask strangers for free stuff, and actually get it?

The larger goal here, of course, is to save energy. Every retail item you purchase has already cost energy—and produced CO_2—to grow, mine, manufacture, alter, or build, and then transport. This energy is unrecoverable—i.e., wasted totally—unless the stuff is reused as is (for maximum return on energy investment), reused as something else (called "new life reuse"), or recycled (for a lower return). Increasingly, economists and environmentalists are advocating a tax to reflect the environmental cost of a product's manufacture and disposal, to encourage reuse.

Sites like www.clothingswap.org can help keep the wardrobe closet under control. No one should buy new clothes that will be used for less than three months—in short, everything infants, toddlers, and fashionistas wear. Check out www.swapstyle.com to tap into the local community—not just for skirts and blouses, but cosmetics and purses as well. You can use the extra cash to sweep an auction of baby clothes and nursery items on eBay, then give them away on Craigslist when your kid has outgrown them. You may never need to waste a day at the mall again.

$5,000–$7,000 is the average amount an American family spends in the first year of a baby's life. (Collectively, it's about $15 billion annually for maternity and baby items.)

20 | Put on a Sweater

If one million people put on sweaters, former President Jimmy Carter would be very happy.

COST	TIME	EFFORT	IMPACT
$0	5 min		

STEP 1

STEP 2

STEP 3

STEP 4

Place your head through the neck hole. Be careful. If you accidentally enter a sleeve, you could be seriously injured.

Take care of any important last-minute errands. Your hands will be out of commission for the next few seconds.

Proceed slowly. Rapid movements can result in static buildup. Use your nose or teeth to guide you and control your speed.

Your midriff is a major source of heat loss, so make sure to keep it covered. Don't forget to button up your pants. You look great!

We lose 90% of our body heat through our skin, mostly from our heads. So if you are still cold, you might try a hat and scarf.

The social benefits of wearing sweaters are obvious—fame, wealth, hiding your spare tire—but what about the environmental ones?

By donning a sweater indoors and lowering your thermostat 2°F, you can save up to 4% on your energy bill and prevent 500 pounds of CO_2 from entering the atmosphere over the course of one year. The possibilities are limited only by the number of sweaters you can purchase or borrow and by your tolerance for immobility. (Caution: turtlenecks do not layer well.)

Visit www.doityourself.com/stry/thermostats for more information

	COST	TIME	EFFORT	IMPACT
If one million people worked at home full-time, we'd eliminate 3 million tons of CO_2 a year.	$0	8 hrs/ day		

Nine out of 10 of us drive to work, with the average commute now up to 25 minutes each way and still climbing. That's a lot of time wasted—and a lot of gasoline becoming CO_2 emissions.

2 weeks

would be saved each year if you eliminated your daily 40-minute commute.

Why not commute to your home office instead? You'll save time and money, increase productivity, and spare the rest of us the emissions.

Working at home is on the rise—in the U.S., it's growing by about 7.5% each year. Employees who work at home have reduced absenteeism, save their companies money in real estate costs, and never ever eat your leftovers in the office fridge.

Don't think it's a day of napping, though: it takes organization to stay focused on the job. Be sure to plan your day like one at the office, with specific goals and a lunch break. Steer clear of the TV, chores, phone calls to friends, and other distractions. You're setting a brave example here, not catching up on E!

22 | Audit Your Garbage

If one million people recycled their metal, plastic, glass, and newspaper, CO_2 emissions would be reduced by 211,000 tons.

COST	TIME	EFFORT	IMPACT
$0	1+ hrs		

$$$

Metal mining consumes 7 to 10% of our global energy output.

They don't teach this at the IRS, but one of the best ways to understand your environmental balance sheet is with a little personal Dumpster diving. Everything in your trash can has a carbon cost, but the messy details are usually out of sight.

You need to see it all together, in the plain light of day, to grasp your excess. Yes, it's disgusting, but so is half the stuff you see on TV. So grab a scale and take out the trash.

HOW TO GET CONTROL OF YOUR GARBAGE

ASSEMBLE your materials: your trash, some old clothes, a tarp, thick rubber gloves (think biosafety), about six cardboard boxes, a scale, a pencil, paper, and a calculator. Find a sheltered area to prevent wind drift.

DUMP trash onto the tarp and sort by material: paper and cardboard, yard waste, food scraps, plastics, metals, glass, and other. One-third is probably paper, while food scraps and plastics each make up another tenth. Only metal cans and yard waste are recycled more than half of the time.

ADD trash from your car, waste from the garbage disposal, and food and beverage waste from work.

TALLY IT UP Use a scale and your estimating skills to give numbers to the garbage that is yours. Weigh your trash from one day. Multiply that by 365. Add an estimate of seasonal trash—wrapping paper, yard waste, Halloween and Christmas decorations—and total it up. The average American generates 1,460 pounds a year.

BONUS POINTS Multiply your count by 30 to reflect the trash behind the trash. American consumers throw away 245.7 million tons, but industry dumps a staggering 7.6 billion tons a year.

PLAN for the future by reviewing each item in your trash. Was it needed in the first place? Did you use it up before you threw it out? Remember that recycled goods take less energy to finish into final products. Think about what you can do to improve your recycling and consumption rates, and do it!

People discard scary stuff—broken bottles, syringes, last month's fish taco—so don't try this anywhere but at home with your own garbage.

96 billion

pounds—over 25%—of America's food goes to waste each year.

The annual value of aluminum not recycled in the U.S. is $800 million.

	COST	TIME	EFFORT	IMPACT
If one million people reused or upgraded their computers, we would save 265,000 tons of fossil fuels.	$0+	1 hr		

$$$

A cell phone is about 14% copper by weight and also contains gold, silver, and about one cent's worth of platinum. These are precious metals, so don't throw away your old cell. Phone and office-supply shops usually have phone-recycling bins.

When it comes to big energy sucks, the usual suspects are obvious and easy to finger—cars, refrigerators, incandescent lights. But there is an inconspicuous environmental Mata Hari. It's sleek, it's small, and it doesn't have a tailpipe: the personal computer.

Like Denmark, *edamame,* and brown shoes, computers just *feel* eco-friendly. But they're actually furtive CO_2 offenders—in ways you might not expect. It's not the plug: electricity consumption accounts for just one-fifth of a computer's lifetime energy toll. The rest comes largely from the tremendous amounts of fuel—usually coal—burned to power its manufacture. Your friendly laptop is about five times more energy-expensive to make than a car or fridge. Producing a two-gram microchip creates 2,000 times that amount of CO_2.

Don't rush out and buy an abacus just yet. PCs are here to stay, and they enable many other environmentally friendly practices like freecycling and telecommuting. (See Skills #19 and 21.) Besides, computers make modern life bearable; how else could we Google our ex...umm...friends?

Because making microchips uses so much energy—and because there's a witches' brew of chemicals and heavy metals in that desktop—it's important to think twice before you dump your old machine for a new one.

WHAT TO DO WITH YOUR OLD PC

UPGRADE A good choice, especially if all you need is a bit more memory or a faster processor.

REUSE Give your PC to someone else—a friend or a charity. Be sure to erase your data, but try to find the original software and licenses. Your donation may get you a tax credit.

RECYCLE Apple, Dell, HP, Sony, and other manufacturers will take care of recycling for you when you buy a new machine, and even salvage what they can from what you send in. For older and off-brand equipment, check with your local government for e-recycling drives or drop-off points. Try to find out where the computers end up—some unscrupulous outfits just ship the old machines to China to be buried.

24 | Convince a Skeptic

	COST	TIME	EFFORT	IMPACT
If one million people convinced their friends, and those people convinced one million friends, and those people convinced one million friends...	$0	1+ hr		

Skeptic

Response

This is never an easy conversation, but don't be intimidated. First of all, you'll never have to debate a "professional" global warming skeptic. Those for-hire scientists are kept on lavish oil rigs in the Gulf of Mexico, where energy executives shower them with gifts of exotic perfume and rare fruit.

Instead, most skeptics you encounter will be people you know well, perhaps even members of your own family. The good news is, their objections are easy to demolish, as demonstrated here.

Be patient, really listen to where they're coming from, and—above all—remember: you're absolutely right.

Global warming?! Scientists can't even agree on whether warming exists, or how it might work if it does.

Virtually every credible scientific organization and study has concluded that the Earth is heating up, and that higher CO_2 levels affect global temperatures.

All this so-called "proof" comes from computer programs and hypothetical projections. There's no real-world evidence of current climate change.

Climate scientists have been keeping accurate records of surface temperature for the past 150 years, and those records definitively show that the Earth is getting hotter—a conclusion that's supported by data gathered from ice cores, satellites, and weather balloons.

Then how come the Antarctic and Greenland ice sheets are getting bigger? If global warming were real, they'd both be shrinking.

Actually, overall, both are shrinking. Greenland's ice sheet loses 48 cubic miles of ice per year. The amount of ice in certain areas may be increasing, but that's because global warming leads to global moistening, and the extra precipitation freezes and becomes ice where it falls. The bigger point: global warming is a shift in climate patterns over a span of several decades, not day-to-day change.

Fine, but if global warming leads to milder winters, better farming, and longer summer vacations, why fight it? It sounds pretty good.

Rising sea levels, stronger hurricanes, and more frequent droughts don't sound good to me. The real concern is how quickly we're making the climate change. Rapid shifts don't give species (including Homo sapiens) time to adjust, and we're not only heating up the Earth to the highest temperatures in human history—we're doing it faster than ever before.

But carbon dioxide is a naturally occurring gas. Humans might have nothing to do with the increase.

I've heard that claim too. It was in an oil industry–funded ad that proclaimed, "Carbon dioxide: They call it pollution, we call it life." Each year, the burning of fossil fuels results in more than 24 billion tons of CO_2 emissions worldwide. CO_2 levels are higher today than they've been at any point in the last 650,000 years.

Even if global warming is real, we can't stop it without ruining our economy and cutting millions of jobs.

Recycling is already a $50-billion-a-year industry, and within 10 years solar power is slated to generate $69 billion a year. There are plenty of opportunities for making money while reducing CO_2. The Industrial Revolution got us here; there's no reason that a Green Revolution can't get us out.

25 | Green Your Cube

	COST	TIME	EFFORT	IMPACT
If one million people shut down their office PCs overnight, we would eliminate up to 45,000 tons of CO_2 per year.	$0	15 min		

COMPUTER

The future is now: host a tele-conference with that webcam your computer came with. One cross-country meeting called off could mean one ton of CO_2 kept out of the atmosphere.

GREENERY

Houseplants can help remove toxins from indoor air. Some are effective for removing carbon monoxide (spider plants and peace lilies); others for the formaldehyde often found in adhesives and furnishings (ficus and aloe vera).

CHICKENS

They'll eat leftovers from lunch and office parties—reducing trips to the landfill and methane release. Roosters may be disruptive if you work in the early morning. (Kidding!)

PAPER

Go paperless. PDFs and emails were meant for the screen; keep them there. Paper makes up one-third of all waste—be sure to recycle it. Recycling one ton of office paper saves almost six tons of CO_2 compared to throwing that ton away.

POWER

Some computers are notorious energy hogs. Set your computer to sleep after a few minutes of inactivity and turn on its energy-saving settings—it'll save your company just as much as $100 per year. Finally, shut the thing down when you leave—that screen saver uses just as much energy as your spreadsheets or Quake 3.

STAPLES

We'd save 120 tons of steel (and all the energy needed to make it) if every UK office worker used one less staple a day for a year. Better yet: staple-less staplers. Behold the power of friction!

LIGHTS

Were you born in a barn? Turn off the lights! Timers and motion sensors help keep things low-energy in a big office.

WINDOWS

If there are operable windows in your office, use them. It'll bring fresh air in—and take a load off the A/C.

MUG

Bring a cup to work. Skip the paper and Styrofoam cups—and the water bottle. (2.7 million tons of plastic are used to bottle water each year, and more than 86% of plastic water bottles end up in landfills.)

BIKE

Hop your bike to work, and talk to your boss about finding a secure place to stash it.

26 | Adopt a Glacier

If all of Greenland's ice sheet and glaciers melt, sea levels could rise 23 feet.

COST	TIME	EFFORT	IMPACT
$0	30 min		

Glacier Adoptio[n]
Certificate

You are the official and
proud parent of the:
Breiðamerkurjókull Glacier

Your love and care will prevent further
recession and maintain
a healthy global balance.

Personalize the plight of glaciers by adopting one. Over the coming years, you'll thrill to see your chosen glacier grow and thrive—or, more likely, shrink and wither.

Your glacier choices are dazzling, at least for the moment. From its ice-capped poles to its mightiest mountain ranges, the Earth is streaked white and blue with ancient ice. Would that one could safely bring home the snows of Mount Kilimanjaro—expected to vanish by 2020—or one of the 26 remaining namesakes in Montana's Glacier National Park—all of which could be gone by 2030.

Glaciers are canaries in the global warming coal mine, sensitive measures of what's happening to our atmosphere. And glaciers are fracturing, liquifying, and flowing into the oceans at alarming rates—90% of Antarctica's glacial mass is diminishing as you read this. The worldwide glacial retreat threatens to feed the climate crisis, swelling sea levels with meltwater and absorbing the Sun's heat into the ocean, where bright glacial surfaces once deflected it.

A good place to start looking for your new, icy friend is the National Snow and Ice Data Center's photo collection, at www.nsidc.org/data/g00472.html. Cherish your adopted ice floe by posting its picture in a prominent place and by checking on it each year. Monitoring your adoptee will be a gradual process. Start with Google Earth or other online satellite imagery, though these sources may remain unchanged for several years. The U.S. Geological Survey and other scientific organizations should have more current information.

The goal here is personalization. Inform yourself about your adoptee. Glacier experts treat their frosty subjects with affection, as if glaciers were massive animals that crawl and calve and carve. Immerse yourself in glacier-speak, adding seracs, moraines, moulins, ogives, and crevasses to your vocabulary. And remember your friend's contribution to your smooth Swiss vanilla bodywash.

Switzerland's alpine glaciers shrank by 18% between 1985 and 2000.

Receding Glaciers: Furtwängler, Kilimanjaro, Tanzania; Athabasca, Rocky Mountains, Canada; Valdez, Chugach Mountains, Alaska; Qori Kalis, Andes, Peru

27 | Stock the Cellar

	COST	TIME	EFFORT	IMPACT
One million grapes will make about 1,600 bottles of wine.	$10+	15 min		

REYKJAVÍK VINEYARDS

Cooler grape-growing regions may benefit from warming: in the coming years, watch for wines coming out of England, Canada, and even Sweden.

So long, Napa

By 2100, nearly all of California's wine regions may be unusable.

Grapes are sensitive little temperature gauges. Our favorite varieties thrive only in narrow bands of temperatures and have very specific tastes in soil, sunlight, and water.

Indeed, rising temperatures are already changing wines, pushing harvest season earlier, raising alcohol levels, and, in the case of Oregon's pinot noirs, fueling a decade of world-class vintages. As the globe continues to warm, however, the classic wine regions of Tuscany, Bordeaux, and even Napa may get too hot for premium grapes. Look instead toward the next wine-making regions: imagine a Québec Syrah, an Albertan Cava, or a Cape Cod Zin—provided the Cape is not under water.

Visit www.liveinc.org for more information

One million homegrown tomatoes will make about 400,000 servings of marinara sauce.

COST	TIME	EFFORT	IMPACT
$25+	summer months		

Until you've tasted a ripe tomato picked from your own backyard, you might find it hard to believe that growing your own food is the way of the future.

Growing tomatoes isn't only about saving food miles—not to mention the money you spend on mealy produce—it's also about reconnecting with the Earth. Tending a cycle of life in your backyard (or windowsill or roof garden) cannot leave you unmoved. Earth helps tomato, tomato helps human, human helps Earth—that's how it's meant to work. Plus, you won't believe the taste.

	STORE BOUGHT	HOMEGROWN
TRAVEL	Approx. 1,500 food miles	6–50 food feet
ORIGIN	In U.S.: Florida (winter), California (June–Nov.), Mexico, Israel In UK: Spain, Canary Islands, the Netherlands, other European countries, Morocco	A 6-inch windowsill pot, a roof garden, or a plot of land. Tomatoes can fruit almost anywhere.
TASTE	Minimal. Bred for sturdiness and yield, tomato crops are picked green, turned red by ethylene gas, and refrigerated for shipping. (You should never refrigerate a tomato.) In the U.S., 81% of tomato varieties have been lost in the past century.	Great! Try growing Miracle Sweet, Celebrity, or Brandywine varieties outside; Pixie, Patio, Toy Boy, or Small Fry indoors and over winter.

29 | Count Your Food Miles

	COST	TIME	EFFORT	IMPACT
If one million people switched to locally produced food for a year, we'd eliminate up to 625,000 tons of CO₂.	$0	5 min		

"Food mile" refers to the total distance food travels in order to be eaten.

Can you imagine the nightmare if every time you got hungry you needed to travel to the other side of the planet before you could sit down to eat?

As absurd as that sounds, we're not bothered by the fact that most of our food has to make this same journey to get to us. To reach your plate, a typical meal travels roughly 22,000 miles. Each ingredient—meat, fish, vegetables, spices, eggs, milk, fruit, grains—is trucked or flown from where it's grown to where it's packaged to where it's sold, burning up fuel. Add the greenhouse gases created by the production and distribution of fertilizers, antibiotics, pesticides, and so on—not to mention the environmental havoc they wreak—plus the paper, plastic, and aluminum used for packaging, as well as their disposal, and you start to get the big picture.

Why should your food have more frequent-traveler miles than you? Why not eat closer to home? Track your food miles—then reduce them. A "locavore" won't eat anything grown more than 100 miles from the dinner plate. A *real* locavore won't even use a plate made more than 100 miles away. So, before buying the next bag of oranges in midwinter, consider some options from closer to home.

REASONS TO REDUCE FOOD MILES

Gives local farmers and rural economies a fighting chance

Mass farming depends on fossil fuels, depletes soils, and poisons waterways

In mass-produced foods, nothing matters but yield, uniformity, and ability to withstand industrial harvesting

Long-hauled food loses vitamins and gains contaminants

Sprawling production systems leave food vulnerable to agro-terrorism

Easier to avoid salmonella, e. coli, and other hazards

Tastier food!

2,993 MILES

A California lettuce eaten in New York devours 40 times its caloric value in fossil fuel to produce and ship.

989 MILES

Making 10 gallons of Florida OJ requires one gallon of diesel and 220 gallons of water to irrigate, process, and transport to northern states.

3,943 MILES

It costs the equivalent of 11,150 calories of energy to fly a two-pound Hawaiian pineapple to Des Moines.

22,000 MILES

A conventional meal creates 4 to 17 times as much greenhouse gas emissions as a locally sourced one.

30 | Befriend Your Farmer

One million calories of food require some three million calories of fossil energy to produce.

COST	TIME	EFFORT	IMPACT
$.99/lb	20 min		

In the U.S., agriculture is responsible for 7% of annual greenhouse gas emissions.

A few short years ago, farmers seemed on the way out. Even though every kid still grows up playing with plastic tractors and watching happy livestock on children's television, the cool little farms themselves—green fields, red barns, clucking chickens—seem to be disappearing.

Turns out, your local farmers are important global warming fighters, not relics of the past. Farmer-to-consumer markets are cutting out fresh food's middlemen—as well as the emissions created by the packaging, storage, transportation, and chemical needs of mass agricultural production. For global warming fighters, the most offensive element is fossil fuel: big agriculture is dependent not only on gas-hungry equipment and transportation, but also on nitrogen fertilizers derived from natural gas.

At farmer's and green markets around the world, producers bring their goods—from organic vegetables and fruits to farm-made cheese, preserves, and meats—directly to market. Even more convenient are Community Supported Agriculture programs (CSAs), which deliver produce direct from the farm each week. You get the freshest foods, and the farmer gets a secure, regular customer.

Not only does buying direct from farmers put food on the table with the least possible carbon cost, it also supports and strengthens rural economies, protects biodiversity, and fosters a sustainable Earth. Plus it teaches kids that strawberries don't grow in the plastic containers at the big box store.

Organic farming and global warming

A recent study of 23 years' worth of data showed that organically farmed soil works as a powerful carbon sink, with soil carbon levels increasing by 15 to 28%. That translates to as much as 3,670 pounds of CO_2 captured by each acre-foot of organic soil each year. Put another way, 100,000 organic farms will eliminate almost 12 million cars' worth of CO_2 in a year. Industrial farming methods do not sink one ounce of carbon.

31 | Eat Your Veggies

	COST	TIME	EFFORT	IMPACT
One million cows burp about 220 tons of methane per day.	$.99/lb	20 min		

Don't be chicken. Stop being a pig. And don't have a cow. Be the first on your block to cut back on meat.

Soy saves lives

One pound of meat requires eight times as much energy to produce as one pound of veggie protein such as tofu.

Relax. This isn't about eating kale; it's about the *single most effective* thing you can do to reduce your carbon footprint: refusing meat.

Pop quiz: which adds more greenhouse gases to our atmosphere, motorized transportation or livestock? Surprise! It's livestock, at 18% of total emissions. The gases coming from cows' rears are even worse, greenhouse-warming-wise, than ol' CO_2. Enteric fermentation—the ruminant's digestive process—produces flatulence, a.k.a. methane, while manure releases nitrous oxide. Even more emissions come from collateral effects: deforestation for pasture, fertilizers for feed crops, and energy to run meatpacking plants.

A 2006 U.N. report, "Livestock's Long Shadow," calls the livestock sector "one of the top two or three most significant contributors" to global warming. (We get only one-third of our protein from flesh, but devote nearly a third of our planet to raising it.) The good news: the report concludes that the livestock sector's global cooling potential is equally vast if we wean ourselves off meat.

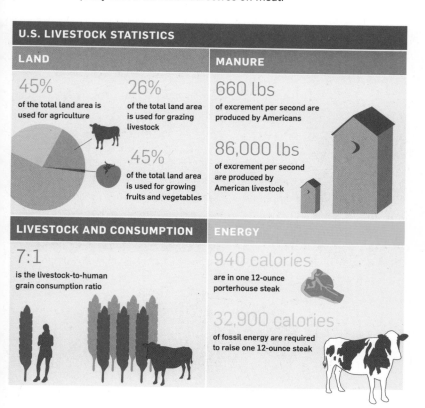

U.S. LIVESTOCK STATISTICS

LAND

45%
of the total land area is used for agriculture

26%
of the total land area is used for grazing livestock

.45%
of the total land area is used for growing fruits and vegetables

MANURE

660 lbs
of excrement per second are produced by Americans

86,000 lbs
of excrement per second are produced by American livestock

LIVESTOCK AND CONSUMPTION

7:1
is the livestock-to-human grain consumption ratio

ENERGY

940 calories
are in one 12-ounce porterhouse steak

32,900 calories
of fossil energy are required to raise one 12-ounce steak

32 | Get Lost in Nature

	COST	TIME	EFFORT	IMPACT
If 1.3 million people visited a U.S. National Park one day in July, it would be just about average.	$0	2+ hrs		

Outdoor ed

Studies have shown that children who play regularly in natural environments are sick less often and show better coordination, balance, and agility.

Remind yourself what it is you're trying to save: nature.

There are still about 1.8 million square miles of unexplored rainforest in the Brazilian Amazon. Nearly one-quarter of the Earth's land is mountains waiting to be climbed. There are almost 600,000 miles of coastline—much of it untouched—and millions of square miles of glorious desert waiting to be crossed. The Grand Canyon continues to take pretty much everyone's breath away.

Wherever you go deep into nature, you'll reacquaint yourself with its spectrum. The color blue—that sky, that water. Also, the greens are nice. Same goes for the purples, yellows, and browns. Ditto the reds and oranges. Oh, and the smells—the air...

There's a beautiful world out there. Hank Thoreau had to write about it. Ansel Adams had to photograph it. John Denver had to sing about it. Yes, nature shows look incredible in HD, but there is nothing in the world like the real thing. Get up off the couch and get out there.

WAYS TO RECONNECT WITH THE OUTSIDE WORLD

WATCH A SUNRISE Don't stare.

EXPLORE A FOREST Hear the crunch underfoot. (Pray it's not a bag of chips.)

CLIMB A VOLCANO Check beforehand that it's dormant and doesn't demand a human sacrifice.

STAND AT THE EDGE OF A LARGE FIELD Especially after a virgin snowfall. Just watch. And listen.

GO FOR A DIVE Or a hike.

TAKE OFF YOUR SHOES And walk on some grass.

One reason to love nature: it's one of the best carbon sinks we have. According to NASA, forests in the U.S., Europe, and Russia socked away more than 700 million metric tons of carbon per year during the 1980s and 1990s.

33 | Harvest the Sun

	COST	TIME	EFFORT	IMPACT
If one million homes switched to solar power, we could reduce CO_2 emissions by 4.3 million tons per year.	$20k+	10+ hrs		

Japan has installed more photovoltaic panels than any other country.

If you think going solar is too complicated or too costly, you must live under a rock. Actually, unless you really do live under a rock, there are plenty of ways to add a little sun power to your life without overtaxing your brain or your bank account.

When residents of Portola Valley, California, near the Stanford University campus in Palo Alto, saw Al Gore's film, *An Inconvenient Truth*, they decided the time had come to go solar. But rather than individually buying and installing photovoltaic (PV) panels on their roofs, they banded together as a community collective—and got themselves a bulk discount of about 30%. Now the owners of 78 homes have enough solar energy to cover about 90% of their household electrical needs every year. That's not just a huge reduction in fossil-fuel dependence, it also means tiny electricity bills—and hundreds to thousands of dollars in savings each year.

The houses are still connected to the power grid, meaning their lights and computers won't go black during a week of summer rain. But the rest of the time, the houses are powered primarily by the Sun.

The Portola Valley homeowner's association president dreamed up the Collective Power Program with a local solar company, Solar City, which now offers similar programs in other California cities and plans to roll out programs in five other states in 2008.

The price of solar systems is coming down, though a set of rooftop panels can still cost thousands of dollars. The next generation of PV will put electricity-generating solar cells in films, which can then be embedded in all sorts of flexible materials. Soon, solar will be embedded in virtually everything, from shingles and windows to the tops of cars, with every bit of solar power displacing electricity produced from fossil fuel.

Solar power's costs are predicted to match coal's by 2010—get your roof ready.

A photovoltaic system that produces 150 kWh every month—about one-sixth of what the average American home uses—keeps 300 pounds of CO_2 out of the air, and saves 150 pounds of coal and 105 gallons of water.

Energy poverty

There are two billion people in the world with no electricity whatsoever, and another one billion who rely exclusively on batteries for their power, and candles and kerosene for light—making conditions for business and education difficult. Solar is one of the easiest, most reliable ways to give power to the one-third of the world's population that has no access to the grid.

34 | Watch the Front Lines

In 2005 the summer Arctic ice sheet shrank to two million square miles—and it may disappear by 2040.

COST	TIME	EFFORT	IMPACT
$0	30 min		

Coral update

By 2050, if we don't control our CO_2 emissions, 97% of the Great Barrier Reef's coral will be bleached every year. The Australian government's ReefTemp offers daily updates on coral bleaching via Google Earth.

Even though you may not have noticed the most recent heat spike—a .3°F increase spread over the past decade—you can still see its effects.

Climate change happens too gradually for us to feel the difference between one day and the next, but that doesn't mean we can't monitor the front lines of global warming.

Landscapes around the world are already decaying, and among this first wave of casualties are some of the most cherished wonders. Keep watch on these fragile corners of the Earth to follow the progress of climate change.

The titan of the bunch, Australia's Great Barrier Reef, is already staggering under the burdens of climate change. The reef's more than 400 types of coral are most threatened. A temperature rise of even 1°F can drive out the algae reef that corals depend on—bleaching them and killing some. In 2002 a "Mass Bleaching Event"—the worst in history—affected at least 60% of the reefs.

Places to watch

1. **New Orleans** Katrina was America's first major battle in the war against warming. Obviously, we lost.

2. **Canadian Arctic** Some 14% of the planet's carbon is locked in permafrost. As the frozen soil thaws, large quantities of CO_2 and methane are released, fueling even more warming.

3. **Venice** Is any city more poignant, and more floodable, than La Serenissima, drowned in tourists all summer and in lagoon waters three dozen times a year?

4. **Komodo National Park** The only wild habitat of the Komodo dragon is in the crosshairs of climate change. Its coral reefs and sea turtles will also be devastated.

5. **Rio Plátano Biosphere Reserve, Honduras** Every tropical rainforest crime in the book—precious trees razed for timber, rare animals hunted, and new species introduced—is causing ecosystem havoc here.

6. **Sundarbans** These islands are home to the largest mangrove forests in the world; 75% of the trees could be destroyed if sea levels rise by just 1.5 feet.

7. **Kilimanjaro National Park** The glaciers here are at least 10,000 years old, but in the last century alone they have lost 80% of their area.

8. **Himalayas** Some glaciers here, which supply up to 75% of summer water for the region, are receding by 50 feet a year.

9. **Amazon** As the Amazon River warms in the next 50 years, up to 60% of the rainforest could become savanna. Rampant logging doesn't help.

10. **Greenland Ice Sheet** Earth's biggest island's ice sheet is disappearing twice as fast as five years ago.

11. **Great Barrier Reef**

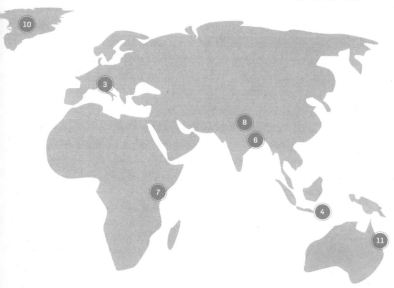

Find out what the Aussie government is doing to protect the Great Barrier Reef at www.gbrmpa.gov.au.

35 | Ride a Bike

If one million people replaced a five-mile car trip once a week with a bike ride, we'd reduce CO_2 emissions by about 100,000 tons per year.

COST	TIME	EFFORT	IMPACT
$50+	30 min		

In Copenhagen, 33% of commuters travel by bike. Perhaps it is no coincidence that a recent survey ranked Denmark as the "happiest" of 178 countries.

2.3 billion

gallons of gas are idled away each year by Americans while stuck in traffic.

About one-quarter of the carbon emissions we produce pours out of the tailpipes of our vehicles. Even if every SUV driver in America switched over to a Prius tomorrow, our cars would still be a leading cause of global warming.

Nature isn't going to wait for all our auto dealers to receive their first batches of ultralight, hydrogen-powered, zero-emission vehicles. Fortunately, you've had the key to solving this problem within your grasp since about the age of eight. It's called the bicycle.

A huge portion of our car travel could easily be done by bike. In California, for example, warm weather makes cycling possible almost year-round, more than half of all commutes are five miles or less, and 40% of all trips are less than two miles. That's easy biking.

Using a bike for transportation will keep you healthy and happy. It will also change the way you see the world. Powered by breakfast rather than a three-dollar gallon of gas, you may come across a friend and have the chance to stop and chat for a minute before continuing on your separate ways. Once you arrive, you'll find a parking spot, oh, anywhere.

While you'll be happy that you've been freed from automotive bondage, you may also notice just how much of your community has been engineered to conform to the needs of cars rather than people. By using your bicycle as transportation, you are asserting a new vision of how the world can be. No need for petitions, sign-waving, or calling on your leaders to do something about global warming. On your bike, you're already doing it.

Life cycle

Half an hour of bicycling daily can increase your life expectancy by up to four years.

Want to make bicycling twice as safe? Bike more—and invite your friends. The rate of bike crashes falls 50% when you triple the number of cyclists on the road.

36 | Decongest Downtown

	COST	TIME	EFFORT	IMPACT
One million commuters waste about 47 million hours per year because of traffic congestion.	$15+	varies		

Urban centers are like the Saudi Arabias of energy efficiency—the world's biggest sources of untapped energy savings.

The densest metropolises are already in many ways the world's "greenest" places, far more efficient in their use of energy, water, and land, and producing far fewer carbon emissions than suburban sprawl. Still, New York and other major U.S. cities could be even *greener*. Single-passenger automobiles aren't the best way to move people into and around dense, crowded streets. Many of the best policies for minimizing car use in cities can also make them into healthier, more pleasant places to live and work.

CONGESTION CHARGING Once suffering from crushing traffic congestion, London now charges drivers £8 per day ($15) to drive into the transit-friendly central business district. All of the revenues generated are poured back into mass transit, bicycle, and pedestrian projects. The same number of people still come into central London but in 500,000 fewer cars per day. Traffic delays are down 22%, and carbon emissions have been cut by 16%. On top of all that, the local economy is doing better. London's mayor is now proposing to increase the charge to £25 per day ($48) for SUVs and other heavy carbon emitters. Brilliant!

CALL-A-BIKE Many German cities have funny-looking red bicycles—called Call-a-Bike's—neatly parked on sidewalks all across town. After preregistering online, all you have to do is pick a bike, call a phone number, and dial in the bike's four-digit code. A cellular signal beams down to unlock it. You pay by the minute, and a half-hour ride costs about the same as the subway or bus. *Wunderbar!*

BUS RAPID TRANSIT BRT—a term for high-quality bus services that use dedicated lanes and other traffic enhancements—has changed the nature of bus riding. Take Paris, where lanes along its most congested avenues have been exclusively dedicated to buses, bikes, and taxis. Linked to a real-time information system that lets riders and dispatchers know exactly when buses will arrive, BRT has turned Parisian buses into fast, efficient, first-class transportation. *Formidable!*

Manhattan story

Eighty-two percent of Manhattan residents get to work by public transit, by bike, or on foot—10 times the U.S. average. Manhattanites consume gasoline at a low rate that the country as a whole hasn't matched since the 1920s and generate less than one-third of the carbon emissions of the average American.

37 | Find a Hero

	COST	TIME	EFFORT	IMPACT
More than 62 million people live in cities that have signed on to hero-mayor Greg Nickels's Climate Protection Agreement.	$0	30 min		

No. 1

is where Atticus Finch, the lawyer in *To Kill a Mockingbird,* ranked in the American Film Institute's list of the top movie heroes of all time (Superman: No. 26).

Heroes save lives. The cape, the mask, the boring alter ego, and the poorly lit underground lair—that's all just icing on the cake. You're not a real hero until you've helped save something or someone.

Luckily for us, real heroes can be found in every corner of the globe, fearless men and women carrying out brave deeds and fighting to save our planet every day. Each has his or her unique powers. What they share are uncommon reserves of superhuman vision, extraordinary stamina, and astounding moral vigor.

ROLL CALL OF ROLE MODELS

JACQUES COUSTEAU

(1910–1997) This passionate Frenchman who (along with Emile Gagnan) brought us a little thing called scuba diving, shared his love of and concern for the life aquatic through fascinating, accessible TV specials. In later life, his energies were spent increasingly on preserving our seas and the life in them.

PETRA KELLY

(1947–1992) Kelly cofounded the German Green Party, the world's first environmentalist party to achieve political significance. While she also advocated for human rights, social justice, and peace, it was the environment that may have benefited most from her attention.

JOHN MUIR

(1838–1914) The Babe Ruth of the ecology movement, Scottish-born Muir—hiker, shepherd, activist, and America's greatest naturalist—cofounded the Sierra Club and helped Yosemite win designation as a national park.

GREG NICKELS

(b. 1955) The Seattle mayor, frustrated with the U.S.'s refusal to ratify the Kyoto Protocol, created the Climate Protection Agreement, which pledges to meet or beat the Kyoto goals. Eight other American mayors initially joined him in taking local action to reduce greenhouse emissions; that number is now more than 450.

WANGARI MAATHAI

(b. 1940) The Green Belt movement that Maathai founded in Kenya in 1977 has planted more than 20 million acres of trees to prevent soil erosion and provide firewood. Though Maathai was jailed more than once for her efforts to end government corruption, in 2002 she was elected to parliament and has served as an environmental minister. She was awarded the Nobel Peace Prize in 2004.

YVON CHOUINARD

(b. 1938) An American-born rock climber and surfer, Chouinard made his own steel climbing spikes as a youth. After discovering that they damaged Yosemite's rocks, he introduced new tools and materials and advocated "clean climbing." He went on to start the clothing company Patagonia, whose initiatives have consistently promoted environmentalism, as well as worker fairness and dignity.

ADAM WAJRAK

(b. 1972) As a young man, Wajrak began writing about animal and plant life in his native Poland, chronicling the disappearance of its ancient woodlands and the grand Vistula River. By his mid-30s, after the fall of communism, he had become his country's leading environmental journalist, working to protect wildlife threatened by post-communist development.

WILLIAM McDONOUGH

(b. 1951) This American architect is a leading voice in environmentally sustainable design and manufacturing. He recently covered the roof of the one-million-square-foot Ford Motor plant in Dearborn, Michigan, with 10 acres of sedum, a ground cover that retains and cleanses rainwater. His book *Cradle to Cradle* has become a bible of the sustainability movement.

38 | Choose the Right Bag

	COST	TIME	EFFORT	IMPACT
If one million people switched to reusable bags, we'd eliminate the need for one billion plastic bags.	$5	5 min		

There are an estimated 500 billion to one trillion new plastic bags used every year. That's as many as two million per minute.

14 million

trees were cut down in 1999 to supply the U.S. with paper bags.

Your hybrid car's waiting in the lot. Your groceries are certified organic and carbon-neutral. You even remembered to pick up a few CFLs for your light fixture. Time to check out. But wait. Paper or plastic? Sometimes, in a carry-out world, there isn't a lesser evil.

The average American family of four tosses out about 1,500 plastic sacks a year, and each one can take up to 1,000 years to decompose. That's quite a legacy. Most of these bags aren't biodegrading; the plastic just breaks up into tinier and tinier bits until it leaches into the soil or water. It's not hard to connect the dots between plastic bags and fossil fuels either. The bags are manufactured from crude oil and natural-gas derivatives, to the tune of 12 million barrels per year.

Paper bags are biodegradable, which counts for a lot, but the energy, chemicals, and water—not to mention trees—consumed in making them amount to an even bigger drain on the environment. And then the bag is used once and thrown away.

So what's the best choice? Let's compare the options.

PAPER OR PLASTIC?

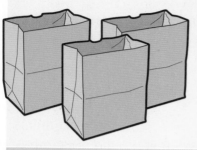

1 ton of bags = 17 trees	1 ton of bags = 11 barrels of crude oil
20% get recycled	1% get recycled
Ingredients: wood, petroleum, and coal	Ingredients: natural gas and petroleum
Could biodegrade in as little as a month, but—due to poor landfill design—actually decompose at about the same rate as plastic	Decompose in 5 to 1,000 years
Each bag leads to 5.75 pounds of air pollution	Each bag results in 1.2 pounds of air pollution; almost 80% less than paper
Generates five times as much solid waste as plastic	40% less energy to manufacture and 91% less energy to recycle than paper
Because of its heft and bulk, it uses more fuel getting trucked to the store	Up to 3% of the world's plastic bags end up as free-floating litter
Its manufacturing process produces 50 times more water pollution than plastic	Easily washes out to sea, where it clogs the stomachs of whales and turtles

THE ANSWER? NEITHER.

You call this a choice? Skip the in-store moral dilemma and Bring Your Own Bag. Whether you prefer hemp (strong and durable), organic canvas or cotton (the touch, the feel), or even nylon (light and cheap), when you B.Y.O.B., you've done the right thing. Better yet, look for a reusable bag made out of recycled materials.

39 | Plant a Tree (Mindfully)

If one million people planted a tree, 1.33 million tons of CO₂ would be absorbed over the trees' lives.

COST	TIME	EFFORT	IMPACT
$10+	1 hr		

CHOOSE

Pick your species. Fast-growing varieties are good, as are long-lived kinds.

DIG

You want a hole about three times the width of the roots.

PLANT

Lift by the root ball! And with your knees.

PACK

Mulch—or organic, worm-made compost—will get your CO₂ fighter off to a good start.

> "The best time to plant a tree was 20 years ago. The next best time is now."
> — Chinese proverb

If you were to pick one simple thing to do to help stop global warming, planting a tree would be the obvious choice. Or would it?

True, a single tree can absorb more than one ton of CO_2 over its lifetime. This is a great thing. Yet a few qualifying points demand attention. It's not that Ronald Reagan was right when he said, "Trees cause more pollution than automobiles do." (Um, no.) It's that when trees decay, they release CO_2 back into the atmosphere (though some remains in the soil). So once a tree has been planted, someone needs to watch over it, and, when the time comes, salvage it for lumber or dispose of it in a landfill rather than mulching or burning it.

If you live in a tropical area, you're in a great position to start planting. Tropical forests filter pollutants quickly and, with their deep roots, draw up water that evaporates into the atmosphere, helping to foster protective cloud cover that reflects sunlight back into space. In temperate parts of the globe, such as the U.S. and Europe, one result climatologists fear is that all those trees you plant will absorb and retain heat from the Sun, contributing to a rise in the temperature of the Earth's surface of up to 10°F by the year 2100 in those regions.

Enough with the bad news. Trees are not just vital for us, of course, but for fungi, insects, and other creatures great and small. Anything that encourages earthlings to appreciate and want to protect nature more rather than less—and looks so pretty in the process—has branches that reach far past tomorrow.

If you can't get to a patch of earth with a sapling, make a contribution to an organization that plants trees. Dig it!

Carbon loans

Trees slowly remove carbon from the atmosphere as they grow and mature. A single tree might absorb one ton of carbon over its 40-year lifespan. When that same tree dies and decomposes naturally—or gets cut down and burned as firewood—it will release most of the carbon it sequestered over the course of its lifetime.

40 | Retrofit Your Career

New investments in renewable energy could create 3.3 million new jobs in the coming decade.

COST	TIME	EFFORT	IMPACT
varies	40+ hrs		

"Green" doesn't mean "antibusiness." It means the planet needs new solutions. New solutions mean new jobs, new companies, and entire new industries—not to mention old industries ripe for reinvention.

After all, someone's going to need to manufacture our clean cars, grow our organic crops, and tweak our new solar panels. We'll need the full spectrum: emission credit brokers, green factory workers, and renewable energy barons. So, whether you start a company or join one, now is the time to get in on the ground floor of the green boom. Here are four ideas to get you thinking.

GET BACK TO THE LAND Farmer's markets are the new village square, and membership in CSAs is on the rise. (See Skill #30.) Many organic farms offer apprenticeships, and even farming neophytes can get a crash course in how a garden grows. Once you've learned the basics, you'll be ready for work in the field or in any number of related jobs, from sourcing local produce for restaurants to teaching good gardening practices.

BECOME AN ENTREPRENEUR Imagine building a manufacturing plant that runs on sustainable principles. At the wind- and methane-powered New Belgium brewery in Colorado, water used to process beer is run through ponds where bacteria eat the organic waste, reducing strain on the city's water-treatment plant and producing methane, which is converted into heat and electricity.

GO INTO SALVAGING Instead of bulldozing an old house to make way for a new one, property owners can hire a local crew of "deconstructionists" to take it down piece by piece—and send the reusable materials to a salvage store. Building-materials salvage yards around the U.S. are filled with all sorts of one-of-a-kind finds.

BUILD THE FUTURE Think you could make everything better? Prove it by choosing a career in green planning or design. Become an environmental engineer and give us biodegradable batteries. Do industrial design and create the first self-disassembling, fully recyclable computer. Or train as an environmental planner and help create your own model city: line the ridges around town with wind farms, end rush hour forever with citywide bike paths, and plop a worm-laden compost heap in the center of the park's new organic farm.

Three other green makers

Terra Plana
A British shoe company that makes eco-friendly footwear featuring nontoxic glues and recycled materials

Simplemente Madera
A Nicaraguan design studio that builds home furnishings out of sustainably harvested tropical hardwoods

La Mediterranea
A Spanish glassware maker that produces plates, glasses, and chic ornamental objects entirely from recycled glass

41 | Invest Wisely

	COST	TIME	EFFORT	IMPACT
One million dollars invested in one socially responsible mutual fund 10 years ago would now be worth $2.2 million. (Your return may vary.)	$1,000+	30 min		

$2.4 billion

in venture capital was invested in clean-energy companies in 2006—more than double the amount invested in 2005.

Not every profit dollar has to come at the environment's expense. It *is* possible to do well by doing good. If you know where to invest, you can save the planet while saving up your nest egg.

Forgive the pun, but global warming stocks are hot. One of the biggest financial trends is finding investments that will prosper in a warming world. If you're a self-directed investor, you can find scores of companies working to help manage the planet's temperature problem—and profit from it. If you like strength in numbers, there are dozens of mutual funds that will put your money in ethical and socially responsible companies—which tend to be the greenest public corporations out there. Other funds concentrate on alternative energy, which many investors see as the industry with the greatest potential in a warming world.

Where to start: www.socialfunds.com is a one-stop clearinghouse for dozens of green and socially responsible funds, from the Appleseed Fund to Walden Social Equity. By investing wisely, you will help support companies that are working to solve global warming. Turn those green values into greenbacks.

42 | Share the Driving

	COST	TIME	EFFORT	IMPACT
If one million people carpooled, 1.7 million tons of CO_2 would be eliminated per year.	$0	30 min		

Feel a twinge of guilt every time you drive solo? It's easy to give your conscience and the planet a break. Share your ride.

The average commuter burns 340 gallons a year, creating a 3.4-ton cloud of CO_2. Ride with one extra passenger and you've cut that figure in half. Find one more and you've cut it by two-thirds.

About 10% of American workers carpool every week. 'Poolers can qualify for discounts of up to 20% on insurance—not to mention drive in carpool lanes—and your employer may offer sweet incentives like free parking, shortened workdays, salary bonuses, and even cash awards.

No car to 'pool in? Try carsharing, where you use a car for only as long as you need, then leave it for the next person to hop into. Carsharing services exist all over the world; a shared car can replace four to eight private autos.

43 | Reuse the News

If we recycle one million more pounds of newspaper, the trees we'd save could absorb 125,000 pounds of CO_2 per year.

COST	TIME	EFFORT	IMPACT
$0	5 min		

67 million
tons of paper are used by Americans each year.

Next time you head to the recycling bin with a big stack of newspapers, remember that even though recycling paper may be the oldest green technology in the world—the Japanese first started repulping their wastepaper in 1031—it's still not the most efficient.

Paper can only be recycled three to five times before its fibers break down; each trip to the pulper only delays its ultimate date with a landfill. Because of the unpleasant fact that paper products are the number-one thing we throw away, it's time to start thinking about how to get more uses out of them. The goal: slow down the fiber flow. Haste really does create waste.

HERE'S A START

PROTECT FRAGILE PACKAGES Instead of cushioning your next shipment of collectible shot glasses in bubble wrap, which is petroleum-based, or Styrofoam packing peanuts—they can languish in landfills for hundreds of years—use bunched-up balls of news-print instead.

MAKE A PIÑATA FOR YOUR NEXT PARTY Soak old newspapers in a flour-and-water paste and form into the shape of your choice. Instead of beating an innocent donkey to smithereens to get the candy inside, choose something guilt-free to smash, like an effigy of Big Oil or a scale model of the Hummer H1.

STUFF IT IN YOUR SHOES TO KEEP THEIR SHAPE It's ecologically better than using shoe trees made out of wood. Newspaper is also a deodorizer.

CLEAN UP AFTER YOUR PETS Line the bottom of kitty's litter box with a bad day's market report. You can save the letters to the editor for Buster.

BURY IT Layered newspaper, covered over with mulch, slowly converts a patch of crabgrass into a bed of prime potting soil. For best results, water down the mulch and then let everything sit for a few months. When you return, you'll have a perfect place to plant your geraniums. You can also add bits of shredded newspaper to your compost heap. Tear up the columnists you hate most and sprinkle them over some eggshells.

USE IT AS WRAPPING PAPER But remember: horrify-ing reports of flood, famine, and drought go on the inside; glamorous photos of celebrity couples go on the outside.

That foul stench wafting from the dump isn't just watermelon rinds and coffee grounds. Rotting paper stinks too.

44 | Build a Bat House

More than one million bats patrol the skies of Austin, Texas, during summer nights.

COST	TIME	EFFORT	IMPACT
$30–$100+	1 hr		

Scientists say that global warming is already a factor in the spread of the West Nile virus. U.S. cases of the disease have risen from 62 in 1999, when it was first discovered, to more than 4,000 in 2006.

Holy climate change, Batman! Dengue fever in Fargo? Malaria in Manchester? Scary thoughts, huh? This may be a slight exaggeration, but global warming is causing the redistribution of tropical pests and their diseases to higher latitudes and elevations, places where these afflictions didn't exist a few decades ago.

Warmer temperatures give mosquitoes and ticks a longer season and extended range. But not only are these pests finding higher latitudes hospitable, they may be experiencing an increase in reproduction rates.

In 2000 the World Health Organization estimated that more than 150,000 deaths could be attributed to disease outbreaks and other conditions sparked by climate change. By 2030, it warns, that number could double.

Barring a custom, head-to-toe suit of mosquito netting, what to do? One answer: get over your fear of bats.

Our maligned friends like nothing more than to swoop down and enjoy a disease-vector snack. Little brown and big brown bats, the most common in North America, can

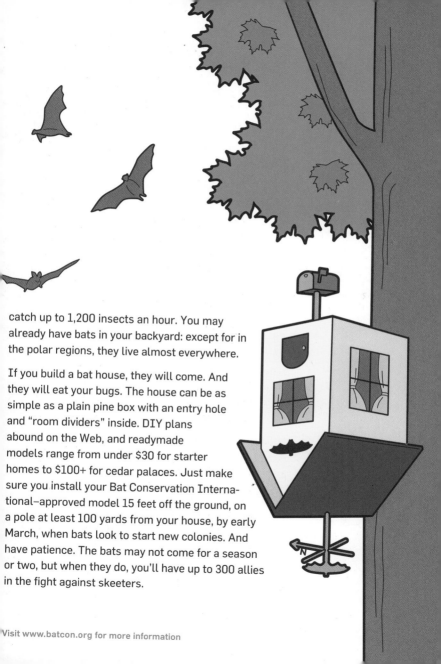

catch up to 1,200 insects an hour. You may already have bats in your backyard: except for in the polar regions, they live almost everywhere.

If you build a bat house, they will come. And they will eat your bugs. The house can be as simple as a plain pine box with an entry hole and "room dividers" inside. DIY plans abound on the Web, and readymade models range from under $30 for starter homes to $100+ for cedar palaces. Just make sure you install your Bat Conservation International–approved model 15 feet off the ground, on a pole at least 100 yards from your house, by early March, when bats look to start new colonies. And have patience. The bats may not come for a season or two, but when they do, you'll have up to 300 allies in the fight against skeeters.

	COST	TIME	EFFORT	IMPACT
If one million people took shorter showers, we'd eliminate 175,000 tons of CO_2 per year.	$0	30 min		

Water heating can add 25% to your home-energy use.

Go with the low-flow

Generally, we'd recommend taking a shower beneath a low-flow shower-head instead of drawing a 30-gallon bath. But because there are two of you, we're going to look the other way just this once. Make the most of it.

In a matter of just a few decades, millions of people will experience water shortages as a result of global warming. According to the latest report from the U.N., every continent will suffer, either from too much water or from too little.

Already, warmer ocean temperatures have brought drought conditions to the western U.S., and there are signs that melting glaciers in Washington and vanishing snowpack in parts of the Northwest, Southwest, and Rockies will make freshwater increasingly scarce in the next fifty years.

Now, you could start panicking. But why not look at the bright side? The coming water shortage could be the perfect excuse to share your bath. Do it for the planet.

Before you proceed to break out the bubbly, make sure you've turned down your house's water heater to 120°F. A 10°F reduction will shave 500 pounds off your annual CO_2 output, and you won't notice the difference.

SEVEN STEPS TO BATH NIRVANA

POUR THE WINE How about a bottle of locally produced organic vino? Imported stuff may as well come with a picture of a gas-guzzling barge on the label.

LIGHT THE FIRE Turn the lights down low and fire up a pair of soy candles. Soy candles burn cleaner than paraffin and they're not petroleum-based, so no fossil-fuel regrets.

***J'AIME BEAUCOUP VOTRE SERVIETTE*, OR: "NICE TOWEL, HONEYCAKES"** Set aside a pair of all-organic cotton or bamboo towels. Bamboo is one of the fastest-growing plants (so it doesn't require as much new land) and it produces more oxygen per acre than most trees. (See Skill #50.)

DON'T GET TOO DIRTY Drop in some all-natural, completely biodegradable bubble bath, and get ready to scrub-a-dub-dub with an organically grown loofah or sensual cotton sponge.

LET THE MUSIC PLAY How about some tunes that are off the hook and off the grid? No need for electricity when you have a solar-powered radio, all charged up and ready to go (not unlike yourself). Try the Freeplay, available at store.sundancesolar.com.

LATHER UP Take a nice, hot (but not too hot!) bath with someone you love. And, if you can't bathe the one you love, it's all right—bathe the one you're with.

Where does the water go?
(in gallons)

330 Watering the lawn for an hour

20 Using water in buckets to wash car

100-125 Washing a car with a hose

1.6 Low-vol. toilet flush

3.5 Standard toilet flush

45 A Jacuzzi-style bath

40 Top-loading washer

25 Front-loading washer

30 Standard bath

24 Regular shower, 6 min

10 Low-flow shower, 6 min

46 | Build a Straw Home

If one million households halved their gas-heating bill, 2.75 million tons of CO_2 would be eliminated per year.

COST	TIME	EFFORT	IMPACT
$30+/ sq ft	40+ hrs		

Know your site

Every location has an abundance of free energy to tap into: sunlight, wind, and the earth. (See Skills #33, 47, and 51.) Building a new house in the right place—and facing the right direction—is crucial.

Building your dream house? Ditch the steel and glass, forget the neo-'60s geodesic dome, and get past the Cinderella castle you once drew in your notebook. Instead, think simple. Think organic. Think straw.

Straw has been used in construction for thousands of years, and bales were turned into buildings in the 1800s on the American Great Plains—where there were no trees. The idea of building with straw was rediscovered by eco-conscious architects in the 1980s, and today's straw houses are not just drafty little houses on the prairie. Instead, straw bales have become the new building blocks of superefficient, CO_2 emissions–reducing green homes.

Bales are made with waste straw that would otherwise have been burned (meaning there's a net reduction in CO_2 emissions from the start). But don't worry about fires—or the Big Bad Wolf. Straw bale is more fire-resistant than traditional wood construction.

50%

As much as half of the average home's energy consumption is used for heating and air-conditioning.

The bales are used to fill in a wood house frame, creating two-foot-thick walls. Coated with stucco and plaster, the straw serves as amazing insulation—two to three times as energy-efficient as traditional construction. Combined with other green building tricks (south-facing windows for "passive solar" heating, efficient heating and cooling systems, and double-paned, low-e windows), your energy use can drop by two-thirds.

Home-energy use is responsible for 18% of annual greenhouse-gas emissions in the U.S.

it www.strawbale.com for more information

47 | Install a Windmill

If 100,000 households installed an ample-sized wind turbine, the annual CO_2 reduction would be 900,000 tons.

COST	TIME	EFFORT	IMPACT
$7,000+	24+ hrs		

Many utility companies offer customers the option of paying a little more to support wind generation.

Nobody likes the thudding sound of a power bill meeting your wallet. Good news: freeing yourself from utility serfdom is getting easier and cheaper.

Wind generation is one of the best technologies available for cutting carbon out of your household diet.

You don't need to be a home-power geek to install a windmill—there are home-energy consultants eager to help you choose, install, and maintain your own turbine. With that help, getting spinning can take as little as one day. If you have enough wind (and cash) for a bigger turbine, you might even find yourself selling electricity back to the utility companies.

Your average wind speed should hit 10 miles per hour—that's best achieved by getting above trees and buildings. Output increases by the cube of the wind speed, resulting in dramatic power increases as the wind blows harder. A 16 mph wind will generate almost 50% more electricity than wind blowing 14 mph.

Right now, unless you need to be off the grid, a single-home wind installation may be too costly. But new turbines are coming on the market that increase efficiency and cost less, and small, building-mounted generators are being developed. The dream of many wind-powered homes may be more than hot air.

THINK BIGGER Wind has astonishing potential on a grander scale. With enough utility-scale windmills—ones 35 stories high with blades longer than the wingspan of a 727—North Dakota alone could light up about one-third of the U.S. Wind can't carry our whole load, but it could go a long way toward capping our carbon output. You can help wind go mainstream by signing up for large-scale wind power. (See Skill #16.)

$$$
Wind power has dropped in cost about 90% over the last 20 years.

To ease your mind

Only about one out of every 10,000 bird deaths caused by humans each year is windmill-related.

The power generated by a one-megawatt wind turbine running 20 years = 29,000 tons of coal or 92,000 barrels of oil.

48 | Green Your Roof

If we grew one million green roofs, we could eliminate 595,000 tons of CO_2 per year.

COST	TIME	EFFORT	IMPACT
$1,000+	20+ hrs		

$$$

A green roof can reduce cooling and heating costs up to 50%.

2 million

square feet of vegetated roofs have been or are being installed in Chicago.

Your roof—or, rather, your green roof—should be a key part of your carbon reduction strategy: plant it, water it, enjoy it.

A green roof is just a planted roof—whether it's sod, a vegetable garden, or a complete woodland of trees and flowering shrubs. A well-planted roof absorbs solar radiation and CO_2, decreases storm-water runoff, provides valuable insulation, gives shelter to a few animals who dig the altitude, and gives passengers in low-flying planes an admirable view.

Green roofs are particularly effective in cities, where the combination of buildings absorbing solar heat and the buzzing cauldron of people, cars, and air-conditioners creates a "heat island effect," raising local temps by 10%.

The sun's heat doesn't get trapped under a green roof, as it does with a regular tar roof; instead, a significant portion of the heat energy contributes to plant growth or gets reflected rather than trapped in the city. As a result, in any locale, green roofs slow heat gain and loss, putting less "load" on your heating and cooling systems—and less load requires less energy and results in fewer CO_2 emissions.

Green roofs can take a thousand different forms, but their structure is basic: a waterproof layer over a standard roof, topped with drainage materials, soil, and, finally, the plantings, which vary from region to region. (New, modular systems let you create a green roof from premade blocks, greatly simplifying construction.) Your green roof might be a thin layer of grass, or an ornate, Dr. Seuss–quality extravaganza. (You'll want to make sure your roof can handle the weight.) A landscape architect can help make your vision of an urban jungle come true.

Can't ante up for a green roof? Paint a dark roof white or silver instead. Black surfaces in the sun can be up to 70°F hotter than white or silver.

The largest green roof in the U.S. covers 10.4 acres atop the Ford River Rouge plant in Detroit.

	COST	TIME	EFFORT	IMPACT
If we skipped one million bottles of water, we'd eliminate 18.3 tons of CO_2 emissions.	$5+	15 min		

Worldwide, we drink some 41 billion gallons of bottled water a year—even though it's often no safer or better-tasting than tap, and wastes fuel and water for bottling and transport. Drink smart: refill a durable, transportable bottle.

A useful rule for reducing global warming: buy less stuff. The corollary: if buy stuff you must, restock with planet-friendly goods.

✓ **RECYCLED PAPER PRODUCTS** One ton of recycled paper uses 17 fewer trees and 67% less energy than nonrecycled paper. Make the full commitment: buy recycled napkins, tissues, paper towels, printer paper, and—yes—toilet paper. (Trust us, the new kinds are actually soft.)

✓ **VEGETABLE-BASED, BIODEGRADABLE CLEANING STUFF** If every American household used just one box of an eco-friendly powdered detergent instead of the petroleum-based kind, we'd save 217,000 barrels of oil a year—that's about 90,000 tons of CO_2.

✓ **FAIR-TRADE, SHADE-GROWN COFFEE** Coffee grown in shade—the traditional way—preserves ecosystems that sequester some of the CO_2 that is produced in making your morning pick-me-up.

✓ **COMPACT FLUORESCENT BULBS** Better, longer-lasting. Need we say more? (Yes? See Skill #3.)

✓ **PERSONAL-CARE PRODUCTS** There's fossil fuel hidden in staples such as lip balm and toothpaste. Look for organic or natural ingredient–only alternatives.

✓ **RECHARGEABLE BATTERIES** Instead of adding toxic chemicals to landfills, switch to rechargeable NiMH batteries. A single charge can last you three or four times as long as one traditional alkaline, and NiMHs can be reused hundreds of times before they finally kick out. When they do die, take them to a local battery recycling drop-off point. Find one within five miles at www.rbrc.org/call2recycle/dropoff.

	COST	TIME	EFFORT	IMPACT
If we planted one million acres of bamboo, we would eliminate up to 4.8 million tons of CO_2 per year.	$8+	15+ min		

Bamboo and trees both sequester CO_2 in their roots and branches as they grow. But bamboo stores more CO_2 and generates 35% more oxygen than an equivalent stand of trees.

Bamboo is a grass, not a tree, and thus grows very quickly, sometimes two to three feet a day. It is during growth that bamboo sequesters the most CO_2.

Bamboo doesn't require fertilizer or pesticides, it stabilizes the soil, and when used in place of wood, it reduces rainforest deforestation. Perhaps most important, bamboo is a versatile industrial material. Here are just a few of the bamboo products that are available.

Choose sustainably grown bamboo products, produced without pesticides and not in newly cleared forests. If you're buying flooring, you can ask for formaldehyde-free materials.

More than one billion people live in bamboo houses.

FLOORING Bamboo flooring is a durable and cost-effective alternative to hardwood.

UTENSILS Because it resists water better than wood, bamboo doesn't warp or split as easily, so it's great for cutting boards, spoons, and bowls.

TEXTILES Bamboo sheets and towels are the latest in luxury linens. The plant's fibers make extremely soft sheets and absorbent towels, as well as fabrics. (See Skill #59.)

Visit www.inbar.int for more information

51 | Dig a Very Deep Hole

If one million homes used a geo-thermal heating system, they'd eliminate 4.4 million tons of CO_2 emissions per year.

COST	TIME	EFFORT	IMPACT
$7,500+	1 wk		

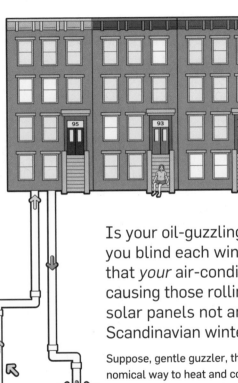

Is your oil-guzzling furnace robbing you blind each winter? Do you fear that *your* air-conditioner may be causing those rolling blackouts? Are solar panels not an option at your Scandinavian winter retreat?

Suppose, gentle guzzler, there was a natural and economical way to heat and cool your home that reduced overall emissions by at least 40% compared to natural gas and central air. And suppose, furthermore, that this method used natural resources found right in your backyard, just below your feet.

It's a real home-heating option called a geothermal energy system, and it puts the Earth's natural heat to work for you. Geothermal systems exploit the Earth's temperature a few feet below the ground surface, which remains relatively constant year-round—between 45 and 75°F depending on the latitude.

Here's how it works: a geothermal heat pump brings water (or water and antifreeze) up through pipes sunk into the ground. In the winter, when the ambient air is cooler, the water absorbs the Earth's heat, which is then concentrated by unobtrusive in-house equipment to warm your environment. In summer, the system acts as a heat sink, taking heat from your home's ambient air into the cooler ground. You need only add energy to power a compressor and heat exchanger.

The pipe network is either looped shallowly beneath a patch of ground—if the property is large—or sunk hundreds of feet into a well or tunnel to reach sufficiently cool ground. (The geothermal system powering one 25-foot-wide apartment building in Manhattan required a column dug more than 1,100 feet into the ground.)

Yes, there is a catch. Geothermal systems can be pricey to install. But some costs can be mitigated by low-interest home-improvement loans, government-sponsored incentive programs, energy rebates, and tax deductions. So do a little research, because this is an amazing heating and cooling alternative for home-owners willing to invest now to save money later.

In Iceland, nearly 90% of all homes are heated by geothermal energy. Because the average annual temperature is just under 40°F, most of this power goes toward keeping Icelanders cozy inside their buildings.

52 | Invent the Antidote

	COST	TIME	EFFORT	IMPACT
Millions of tons of sulfur were sent into the atmosphere by Mount Pinatubo's eruption in 1991; the Earth's surface cooled by .9°F in the following year.	genius, sweat	1 yr+		

If Thomas Edison knew how much his lightbulb would someday contribute to global warming, he probably would have kept on going until he invented a solution for carbon emissions. Alas, he was an inventor, not a soothsayer. That leaves you, ingenious reader, to solve humanity's biggest challenge.

The good news is that you won't be tinkering alone. The world's scientists, engineers, and lay geniuses are already hard at work in dozens of fields, from clean energy to large-scale carbon sinks to electric vehicles. A key culprit is cement, the world's most commonly used manufactured material. At least 5% of man-made CO_2

emissions come from the production of the cement that goes into making most concrete. Hopeful glimmer: an Australian inventor has devised a cement that actually absorbs CO_2—a ton of cement sinks a half-ton of it in one year—can be recycled, and uses waste materials.

Where else might a budding Edison focus his skills? Consider biofuels. Bio-diesel production usually requires intense heat to kick-start the reactions that turn feedstock into fuel. Reducing the heat necessary—perhaps adding an enzyme here and a microorganism there—could flip the equations strongly toward biofuels.

Or batteries. What electric cars need is a smaller, cheaper, less toxic, longer-lasting, and faster-charging kind of battery. While you're at it, get cold fusion ready for the energy market and you're sure to get a Nobel.

But accolades aren't the only thing at stake. The people who just awarded $10 million for the first civilian space flight have put up another $10 million to turn an urban legend—the 100 mpg car—into reality. It's called the automotive X Prize, and the winning vehicle must be market-viable.

You could bank another $25 million via the Virgin Earth Challenge, offered by megapreneur Richard Branson. To win it you'll need to engineer the net removal of greenhouse gases from the atmosphere each year for 10 years—no "countervailing harmful effects" allowed.

Cash money may move mountains, but keep in mind that Toyota spent around $1 billion developing the Prius, which doesn't even make it halfway to the X Prize's mileage benchmark. The real prize here is in market share. The CO_2-beating technologies of the future will remodel the world economy, as surely as the car and the computer did in the last century.

High-tech ideas that might pull us out of our CO_2 mess

Sunscreen
High-flying balloons release sulfur particles in the atmosphere, reflecting sunlight and heat back into space.

Space mirrors
Giant mirrors assembled in space— tens of thousands of them—angle away the Sun's rays.

Sink
Seed the ocean with iron to create a phytoplankton baby boom; the critters draw excess CO_2 from the atmosphere.

Sequester
Convert CO_2 to an icy slurry deep in ocean trenches.

Visit www.auto.xprize.org and www.virginearth.com for more information

53 | Skate on Old Tires

One million tires require the energy of 7 million gallons of gas to produce.

COST	TIME	EFFORT	IMPACT
$0	weeks		

As many as three billion scrap tires wallow in piles around the country. Tire dumps can contaminate surface-water runoff. They're fire hazards that can self-combust and take months to burn out, all the while emitting toxic smoke. They are breeding grounds for rodents and insects like the disease-bearing Asian tiger mosquito. Forty-six states have banned whole tires from landfills; many tires are disposed of illegally, polluting woods, deserts, and empty lots.

290 million

tires are scrapped annually in the U.S.

Scrap tires are a particularly fossil-fuel intensive type of waste, not to mention a potential breeding ground for dengue fever and encephalitis. Rather than dumping them in the forest, incorporate used tires and extreme sports into your survival plans by building skate parks with tire foundations.

Skate parks are hard to argue with: they provide good, clean fun, as well as a place where kids (and adults) can participate in a constructive activity instead of running the air-conditioner. By employing the "Earthship" technique (a popular term for utilizing old tires to make rammed-earth houses), you can lend skating an edge of ecological righteousness. In Washington, D.C., the nonprofit East Coast Round Wall Foundation recovered tires from abandoned lots and National Parks, filled them with dirt, and laid them down as the foundation of the Green Skate Lab. After forming the walls of the bowl with tires, workers put down rebar and then poured concrete.

Visit www.skateparkguide.com for more information

The result: a 100% volunteer-built skate environment made entirely of reused and recycled materials.

Skateboarders are known for reanimating discarded junk into structures built for carving and grinding. In Portland, Oregon, at Burnside Park, donated and scavenged concrete brought a crime-ridden space back to life as a skate park. At the Chehalem Skate Park in Newberg, Oregon, a truck axle and hub turned on its side was repurposed as a spinning yellow obstacle. The practice of salvaging discarded wood, busted bricks, automobile carcasses, and other industrial detritus is ingrained in the skater ethos.

Take a cue from these urban archeologists: gather some salvageable materials and build a play area that is safe and useful. Recycling doesn't need to be lame.

Nearly three million Americans live within 10 miles of a nuclear power plant.

COST	TIME	EFFORT	IMPACT
$0	15 min		

First things first: "nuclear" is not pronounced "*new-cue-ler*." Or "*new-cler*." Since it's likely to be rolling off more tongues than ever in the coming years, it's best to get that straightened out right now.

Long the scourge of environmentalists who associate nuke plants with nightmare scenarios (backed up by real-life events at Three Mile Island and Chernobyl), this atomic power source has lately been getting a second look from some of its former enemies, who wonder if it might be a short-term solution to the problems of dwindling oil supplies and global warming. Be wary of the hype.

At first glance, nuclear's appeal as a tool to stop global warming may appear to be simple and persuasive: it's not a fossil fuel and it's carbon-free—if you don't count the uranium mining and reactor building—making it a possible method for controlling CO_2 emissions.

But there are still a few kinks to be worked out: potential plant meltdown, where to store the radioactive waste (which takes 10,000 years or more to decay), plutonium falling into terrorist hands. Plus there's the fact that uranium is a nonrenewable resource that is costly and

Nuclear costs

Trying to safely manage and store spent nuclear fuel accounts for about 5% of the total cost of generating electricity from nuclear power.

environmentally damaging to extract from the earth. (Estimates of uranium reserves vary.)

Researchers at the Natural Resources Defense Council (NRDC) calculated that in order for nuclear energy to displace enough fossil-fuel energy to make any real difference to global warming, worldwide nuclear output would have to double by 2050 and continue at that capacity for at least 50 years. Yet even that massive increase in atomic energy, the NRDC says, would only prevent .33°F of global temperature rise.

Of course, one-third of a degree is better than nothing. But it's far from a "glowing" review. Is nuclear power worth the cost?

Many think not. The jury is still out, but one thing is certain: nuclear power and the debate surrounding it will be with us for years to come.

55 | Choose Your Fuel

	COST	TIME	EFFORT	IMPACT
If one million cars switched to bio-diesel, we'd eliminate 5.2 million tons of CO$_2$ per year.	$3+/gal	5 min		

average gallons of
fuel burned per year
per vehicle

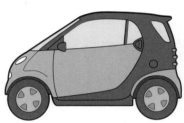

A moment of silence, please, for Cheap Oil. Pretty much everything around you was built upon the geyser of cheap and abundant energy unleashed in the 19th century.

Mad scientists and venture capitalists are busy trying to develop the successor to Cheap Oil. Ethanol from switchgrass and ditch weed, bio-diesel from algae and turkey offal, geothermal hydrogen, tidal electricity, methane ice, P-series fuel—it's enough to give your gas tank a real identity crisis.

Each alternative-fuel option has pros and cons, and environmental impacts vary. For example, biofuels seem carbon neutral: the plants absorb CO$_2$ as they grow and release it when they're burned. But it also takes energy to grow the crops, harvest them, cook them, and distribute the fuel.

If we were really smart, we'd invest as much of our remaining fossil energy as we can in wind turbines and solar panels. The best fuel is the one you do not burn up in the first place.

The right fuel for your time and place may change as technology improves. The race is on!

9,760
pounds of CO_2 are emitted on average each year by a single gas-powered car.

THE SCORE SO FAR

🌽 BIOFUELS *ethanol, bio-diesel, methane*

✓	Practically everything organic—including waste—can be transformed into fuel	✗ ✗	High subsidies
			Creates a fuel-versus-food choice
			High environmental costs of farming

⚡ ELECTRICITY

✓	Centralized generation makes carbon capture easier	✗ ✗	You need to buy a new electric car
	Existing infrastructure		Battery technology limits range and refueling time
	Lots of sustainable sources		Lots of unsustainable sources

H HYDROGEN
1.0079

✓	A new way of looking at energy	✗ ✗	Currently made from natural gas, along with lots of CO_2
	Waste product is water		
	Highly efficient when used in fuel cells		Could be many decades before technology matures
	Carbon neutral if produced from renewable sources		Infrastructure price tag: $500 billion

🦕 DINOFUELS *coal, natural gas, methane, oil*

✓	Huge amounts still available	✗ ✗	Major contributor to global warming
	Massive infrastructure and vehicle base exist		Extraction is messy
			Petro-politics are even messier

56 | Drive a Frybrid

More than one million gallons of fuel could be made from waste vegetable oil collected from San Francisco's 2,600+ restaurants.

COST	TIME	EFFORT	IMPACT
$500+	40+ hrs		

Vegetable oil lubricates better than dinofuel, so the engine lasts longer and runs quieter.

It is a brave new world for your gas tank. There's hydrogen, natural gas, even captured brake-power replenishing hybrid-car batteries. No fuel may seem braver, though, than waste vegetable oil.

If you were the Lego-set type as a kid, you have the skills to convert a diesel car or truck to run on straight waste

vegetable oil (WVO) harvested from local restaurants. The main trick is to heat the grease before injecting it into the engine—a heated tank in your trunk will take care of that. You'll also need to secure a stable supply of used oil—Chinese buffets are a good source—filter it, and get used to a vaguely fried smell following you around.

If running on fry grease is too big a shift, try bio-diesel, a fuel produced from plant and animal products instead of petroleum. Both bio-diesel and vegetable oil are usually thought of as carbon-neutral fuels; although they release carbon when they're burned, that carbon was absorbed by plants, meaning that you are not adding fossil carbon to the air. There is debate on this. Bio-diesel also burns cleaner—almost sulfur-free and with less carbon monoxide.

Bio-diesel can be brewed with everything from canola oil to turkey guts to algae, but the purest, geekiest, most carbon-neutral expression is the grease car running on waste vegetable oil. (Register at www.fillup4free.com to see local sources.)

Any car that runs on diesel fuel can be converted into a grease car (or "frybrid"), but old, built-to-last Mercedes are a popular choice. The had-money-30-years-ago impression doesn't hurt either. Expect to pay at least $2,000 for an old Benz. Veggie-oil conversion kits start around $500. Some shops specialize in grease-car conversions; expect a complete installation to cost at least $1,500.

Be sure to connect with your local bio-diesel scene first; you'll want to line up fuel sources and familiarize yourself with other potential problems. This technology works best for people who like to tinker; expect to spend time under the hood.

3.8 billion

pounds of grease produced at U.S. restaurants each year would make 495 million gallons of bio-diesel, meeting America's driving needs for 1.4 days.

57 | Catch the Rain

	COST	TIME	EFFORT	IMPACT
If one million Americans used rain barrels, they could collect enough water for the daily usage of about one million others.	$50	30 min		

Redoing your driveway? Look into permeable pavement, which filters water into the ground instead of jetting it into the sewer.

Water shortages

1.1 billion people do not have access to safe drinking water.

2.3 billion people are projected to live in countries affected by chronic or recurring shortages of freshwater.

Scientists talk about climate change rather than global warming because rising temperatures will also shift patterns of wind and rain.

While many areas will be drier, including vast regions of Africa and Asia, most of the U.S. is likely to get more rain. Sounds nice, except that many regions can't handle the rain they get already. Storm water washing off roofs and roads crashes into streams with such force that it rips out banks, riverbeds, and infrastructure.

Storm water can do double damage. Instead of soaking into the ground, where it can slowly recharge rivers and lakes, giving life instead of ripping it away, high levels of dirty, high-impact storm water can rush across the surface and spread pollutants. Combined with decreased groundwater, storm water damages the ecosystem and helps account for why more than 40% of American rivers and lakes aren't safe for swimming or fishing.

The problem is likely to get worse as climate change progresses. The solution is delightfully low-tech: slow the water down so that it can soak into the earth, which stores and filters it.

PLANT A RAIN GARDEN If you get a lot of runoff, a rain garden can help tame the flood. (Lawn grasses have very shallow roots, which limit their ability to absorb and filter runoff.) Follow the water during the next storm to find where it collects. Sculpt a shallow depression to capture the water. Plant deep-rooting native perennials and shrubs to help the ground absorb the water quickly. Different plants work best for different regions—a quick Web search, or a call to a local college's agricultural extension agent, will get you a comprehensive list.

RAIN GARDEN PLANTS BY U.S. REGION

NORTHEAST	✓✓	Elderberry, silky dogwood, Joe-pye weed, milkweed
WEST	✓✓	Monkeyflower, cottonwood, huckleberry, Pacific Bleeding Heart
SOUTHEAST	✓✓	Ferns, giant coneflower, dwarf palmetto, possumhaw, bald cypress

INSTALL A RAIN BARREL You'll want a big, clean plastic barrel—a garbage can will do in a pinch, but better to buy a purpose-built model or make one yourself. (Get DIY instructions at www.raingardens.org/docs/rainbarrel.pdf.) Many cities buy barrels in bulk and sell them to residents at a reduced price.

600 gal can be collected from a 1,000-square-foot roof covered with an inch of rainfall.

	COST	TIME	EFFORT	IMPACT
If one million people moved into Energy Star–certified homes, we'd eliminate about 5.6 million tons of CO_2 per year.	$ 0	1 min		

If you want to be a responsible "green" consumer, you have to know what you're buying. That means reading labels. Trouble is, some labels aren't exactly reliable. But knowing which ones are legit is essential to ensure that the actions you're taking will really change the game.

The sources of the eco-labels showing up on so many products can be either governmental or private. Neither is a guarantee of authenticity. Here is a rundown of some of the most common.

ENERGY STAR This cursive logo has been a friend to consumer and industry alike since the EPA approved its first Energy Star product line in 1992. Starting with computers, the label now certifies energy-efficient appliances, cooling systems, and even buildings. **Reliability: Excellent**

FAIR TRADE CERTIFIED More than a force for economic justice for the developing world's small farmers, Fair Trade standards—handed down from an umbrella organization in Bonn, Germany—encourage eco-friendly farming methods and ban a dozen of the most pernicious pesticides. **Reliability: Excellent**

GREEN SEAL Sealing since 1989, this respected standard for environmental responsibility achieves scientific rigor and public transparency by building consensus between stakeholders and industry. Green Seal certifies a wide range of products, from paints to windows to lodgings. **Reliability: Excellent**

LEED A building graced with the coveted LEED (Leadership in Environmental Energy and Design) label conforms to standards reliably applied by the U.S. Green Building Council. The successful LEED labeling formula may soon extend to the healthcare industry. **Reliability: Excellent**

USDA ORGANIC The "organic" label gets slapped on many products that may not truly deserve it. (Let's not even start with "natural" and "healthy" labels.) The USDA's standard for sustainably produced, chemical-free food is one of the best available, but it gets criticized for failing to adequately define just how much ranging a free-range chicken ought to be allowed. **Reliability: Good**

SUSTAINABLE FORESTRY INITIATIVE The standards behind this label—created by the American Forest and Paper Association—are about as strict as Mary Poppins. From lumberyards to paper mills, every certified timberlander is allowed the "prudent" use of forest chemicals like herbicide and 120-acre clearcuts. Wood-focused environmentalists prefer the FSC (Forest Stewardship Council) stamp of approval, which certifies the paper this book is printed on. **Reliability: Poor**

If one million people washed their clothes in cold water instead of hot, we'd eliminate 250,000 tons of CO_2 per year, and keep our colors brighter.

COST	TIME	EFFORT	IMPACT
$20+	30 min		

BEFORE

What happens to the chemicals used to dye that red dress so vibrantly? They seep into the water supply. Choose clothing dyed with all-natural vegetable coloring.

DRY CLEANING
The plastic wrapped over your dry-cleaned jacket isn't the only danger sign—85% of dry cleaners use the chemical PERC, which has been linked to cancer and reproductive damage.

BOOTS
Fonzie may have been cool, but his love affair with leather made Mother Earth cringe. Tannic acid, formaldehyde, and other chemicals are added to hides to prevent them from decaying on your flesh.

FABRIC
Nylon and polyester suck up fossil fuel during their manufacture. These synthetics are a fashion don't—keep an eye out for recycled polyester instead like Patagonia's Capilene, or all-natural weaves.

TUNES
You're jamming to your old-school Discman and suddenly that Air Supply song cuts off! Non-rechargeable batteries are so 1983; if you toss them, you're adding mercury, zinc, lithium, and cadmium to the landfill.

EZ KLEEN

FABRIC
Super soft and draping the body in the most flattering way, bamboo as a shirt fabric not only looks great, but does not require pesticides and is naturally regenerative.

JEWELRY
When you buy handmade jewelry, you're supporting companies that don't add pollutants to the environment in huge, factory-scale quantities.

GREEN SLEEVE
Though coffee sleeves are usually recycled, do 'em one better by bringing your own chic, reusable sleeve.

BELT
Buckling up with a hemp belt makes you look good, and knowing it reduces deforestation will make you feel better too.

JEANS
Denim is made from cotton, the most pesticide-intensive crop there is. Makes you itchy just thinking about it! Keep on the lookout for new organic-cotton denim.

WEDGES
Footwear, leather or synthetic, is often nonbiodegradable. Try investing in shoes that use vegetable-tanned leather or organic materials, like canvas or cotton. And when you've worn them out, try to get them resoled instead of buying a new pair.

OUTFIT
Recycled clothing takes textile waste and energy costs off the table. Vintage clothing, accessories, and particularly bags (the more retro the better) add one-of-a-kind style.

The nine personal-care products the average adult uses contain 126 chemicals.

60 | Enlighten Your Friends

	COST	TIME	EFFORT	IMPACT
One of the Princeton Carbon Mitigation Initiative's "wedges" represents the reduction of one gigaton of carbon over 50 years.	$0	15+ mins		

According to a 2005 survey, 82% of Americans use TV as their primary source of information about environmental problems.

Fighting global warming is not just about changing your own personal energy and consumption habits. It's getting businesses, institutions, and governments to change theirs. If we don't reform on a grand scale, all that weatherstripping at home will be a novelty act.

Your task is to sprinkle some high-concept, low-carbon strategies into everyday conversation. Friends nodding like they already know? It's beginning to work! Repeat as necessary; you'll know your job is done when CEOs and world leaders are tackling the big tasks without your help. Here are the top policy-level issues that keep energy wonks up at night.

THE PRINCETON STABILIZATION WEDGE Big problems require big solutions, and the "Wedge Game" teaches just how big we need to think. It challenges you to stabilize greenhouse-gas emissions over the next 50 years by fitting some of the biggest puzzle pieces together: increased efficiency, fuel switching, carbon capture and storage, nuclear, hydrogen, wind, solar electricity, bio-fuels, and natural carbon sinks.

RAISE CORPORATE AVERAGE FUEL ECONOMY (CAFE) STANDARDS American fuel economy is dismal compared to Europe and Japan, despite the fact that the same car companies compete in each market. The difference is regulatory cowardice. Elect leaders who will put the brakes on gas guzzlers.

RENEWABLE PORTFOLIO STANDARDS

Twenty states have set binding target dates for utilities to get a minimum percentage of their power from renewable sources. Does your state have a minimum standard? Is it high enough? The national percentage is currently 2 to 3%. Twenty percent by 2020, increasing 1% each year thereafter, is aggressive but attainable.

SMART GROWTH

The locations for jobs, shopping, and people—and how they're connected—have a lasting impact on energy use. Advocate for smart growth in your community.

ENERGY-EFFICIENCY RESOURCE STANDARDS

Set conservation and efficiency standards for utilities, obviating the need for hundreds of coal-generating power plants. Improve energy-efficiency standards in appliances and building codes.

CARBON CAPTURE & STORAGE

Also called sequestration, the idea is to take carbon from the atmosphere and store it—usually deep underground. It needs perfecting and can only be done on an industrial scale.

END FOSSIL-FUEL SUBSIDIES

Currently, oil, coal, and natural-gas profiteers pad their bottom lines via tax credits and cheap federal leases.

CONSISTENT INCENTIVES FOR RENEWABLE-ENERGY RESEARCH

Wild swings in federal tax credits for wind power have hampered growth of the industry, while a similar roller-coaster approach has delayed research on cellulosic ethanol. Renewable energy feed-in tariffs require utilities to pay a set price for renewable energy and would help further stabilize the industry.

TAX THE (CARBON) RICH

Economists argue about this all the time, but a number of them now believe we need a significant gas tax increase whose proceeds would be invested in renewable-energy research and transit infrastructure. Another option: an aggressive, economy-wide cap on carbon emissions and the establishment of carbon credit trading, allowing the market to sort out the most cost-effective approach.

VOTE

Eighty-one percent of Americans agree with the statement, "It is my responsibility to help reduce the impacts of global warming." But that opinion only matters if your representatives actually represent you.

	COST	TIME	EFFORT	IMPACT
More than 63 million votes were cast in the 2006 *American Idol* finale. It can't be that hard.	$0	10 min		

Local politicians can make a difference. In the U.S., more than 450 mayors have signed a Climate Protection Agreement, pledging their cities to meet or beat the goals of the Kyoto Protocol. In London, Mayor Ken Livingstone has announced an ambitious program to cut the city's carbon emissions by 60% within 20 years.

It used to seem that if you didn't vote on Election Day, it wasn't the end of the world. Now, it just might be.

Without governments committed to protecting the planet, there's only so much we as individuals can do. What we need are (drumroll)...elected representatives! Hey, it worked wonders for civil rights.

If we've learned anything this century, it's that every vote counts. But for whom? Party labels are not a fool-proof way to tell where a candidate stands on global warming. Some candidates will say the right things at first, but when you ask for details, you find that they're blowing hot air at you. Which is ironic, if you think about it.

1) True or false? You favor either an overall reduction of greenhouse-gas emissions or a slowdown in the rate of their growth.

T   F

2) Do you believe that the U.S. must reduce its global warming emissions by 60 to 80% by 2050?

Y   N

3) True or false? Your idea of recycling is to shred all regulations that protect the environment.

T   F

4) Do you support tax incentives to increase the number of green buildings and homes?

Y   N

5) Do you favor incentives or subsidies that make alternative energy competitive with fossil-fuel energy?

Y   N

6) True or false? The endangered species list should come with some nutritious recipes.

T   F

7) Do you support subsidies for crops that can be turned into biofuel?

Y   N

8) What gas mileage standard should cars meet?
    20 mpg   35 mpg   45 mpg   75 mpg

9) The United States should
   (A) take the lead on international efforts to reduce global warming pollution.
   (B) continue to crap up the planet and let everybody else worry about the consequences.

10) From the following list, circle the three most effective strategies for dealing with global warming:
   (A) Joining international environmental treaties
   (B) Triple-dog-daring polluters to reduce emissions
   (C) Writing legislation to limit CO_2 output
   (D) Waiting for a league of superheroes
   (E) Educating the public
   (F) Handing out free cocktail umbrellas

Vote for the candidate with the most correct responses: (1) T; (2) Y; (3) F; (4) Y; (5) Y; (6) F; (7) Y; (8) 75 mpg; (9) A; (10) A, C, E

Seventy-five percent of Americans believe that government should restrict emissions on cars and factories in order to reduce global warming.

62 | Move to Stockholm

If one million Americans cut their emissions as much as Stockholm has, we'd eliminate about 1.3 million tons of CO_2 per year.

COST	TIME	EFFORT	IMPACT
taxes	1 day		

As if you needed added incentive to move to a place with clean air, near-constant summer sunshine, and beautiful furniture, Stockholm may be Europe's most climate-forward city. Pack up your essentials, slip on some clogs, and join Scandinavia's green revolution.

The Swedish capital, population 750,000, has already cut its greenhouse-gas emissions by 17% since 1990, but those overachieving Swedes keep setting the bar higher. The city now has a plan to be fossil-fuel free by 2050. And if that's not enough, they're anticipating meeting that goal in some sectors—public transportation and heating—30 years ahead of schedule.

Sweden has slashed its use of fossil fuels, from 70% of energy in 1970 to 30% today, and beefed up its reliance on biofuels such as ethanol and biogas. More than 70% of Stockholm's heating is already supplied by biofuels, with one former coal plant converted to run on them entirely. By 2009, all gas stations in the city will be required to sell at least one form of renewable fuel, and by 2020, Stockholm's entire fleet of buses will run on renewables.

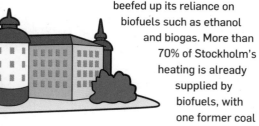

Still in the works are plans to convert all of the city's food waste to biogas that would power municipal vehicles and a system of ferries for the city's waterways. It's not just Stockholm, either: more than 15 years ago, Sweden was one of the first countries in the world to impose a tax on CO_2 emissions, and will be increasing that tax throughout the rest of this decade. Funds from energy taxes made up one-tenth of the country's revenue in 2004.

On second thought, if we all hightailed it over there, the good citizens of Stockholm would never be able to achieve their goals. Here's a better idea: push for progressive policies in your own city.

Cities with Stockholm syndrome

Boulder, Colorado
Voters there recently approved the country's first "climate tax."

Chicago, Illinois
Here is the country's largest network of green roofs, with more than one million acres of building-top gardens.

Copenhagen, Denmark
Sixty-eight percent of commuters here ride their bikes or take public transportation to work, and the city provides free bicycles for getting around downtown.

London, England
The city, which boasts an innovative, energy-efficient city hall, has a many-faceted plan to reduce its CO_2 emissions 60% by 2025.

Portland, Oregon
This longtime leader in sustainability uses methane from its waste-water treatment to generate electricity.

If one million people commuted by train instead of by car, we'd eliminate 1.2 million tons of CO_2 per year.

COST	TIME	EFFORT	IMPACT
$2+	20 min		

1,000 miles

of travel produces how many pounds of CO_2 per passenger?

Mass transit: 260 lbs

Intercity train: 450 lbs

Economy car: 590 lbs

Jet: 970 lbs

SUV: 1,570 lbs

Jets will soon be reviled as atmosphere-killing, CO_2-producing dinosaurs. What's a traveler to do? Jog across the country? Ride a bike to the family wedding in Chicago?

How about the train? Trains are the most ecologically low-impact way to cover long distances. In addition to being more energy efficient than cars and planes, rail is the perfect antidote to suburban sprawl. Trains tend to to organize human settlement patterns around walkable, bikeable, transit-friendly town and city centers, while air and highway systems tend to encourage the opposite.

Only 60 years ago, the U.S. had one of the best intercity rail networks in the world. Jet travel, the interstate highway system, and the exodus to car-oriented suburbs changed all that.

Today, poor, beleaguered Amtrak is kept starved of funds by members of Congress who insist that the nation's rail system turn a profit while they hand over billions of dollars of subsidies to the highways and airlines. Neglect of our nation's rail system has become so severe that passenger trains are nearly useless in many places—forced to stop and sit on a sidetrack while freight cars pass. As a result, Amtrak frequently arrives hours late during long-distance trips.

It doesn't have to be this way. Rail has the potential to bridge the gap to the future. France's TGV recently set a new rail-speed record of 357 miles per hour. Japan is testing nearly frictionless maglev trains that levitate 10 centimeters above the track and reach speeds of 300 mph or more using relatively little energy and no moving parts. In Germany, when your plane lands at Frankfurt International Airport, there's a train every 15 minutes to the central station, where you can catch another train to pretty much anywhere in the country. On board, you can plug in a laptop; the food is good, and there is plenty of space to move around.

The U.S. could have all that too. The train could once again be the fastest, most pleasant, and most environmentally friendly way to travel across the United States. It wouldn't take some technological miracle to make it happen. We have the know-how. We just need the political will. Start with a train trip. It might be great. Or it might suck so badly that you're inspired to do something about it.

Pound for pound, how does the train compare to the jet, in terms of CO_2 emissions? If you include the cab rides to and from the airport, a round-trip flight between Boston and Washington, D.C., generates 776 pounds of CO_2 per traveler. Making the same trip on the Amtrak Metroliner produces only 360 pounds of CO_2.

If one million people used just one set of hotel linens for a week, we'd save 1.5 million gallons of water.

COST	TIME	EFFORT	IMPACT
$0	5 min		

Common hotel-guest crimes

Cranking up the A/C

Leaving on all the lights

Pilfering the toiletries (which forces increased production of them)

Guilt-reducing strategies

For unenlightened hotels, supply your own "save towels and sheets" cards, from greenhotels.com/catalog/traveler.htm.

Quit collecting wee shampoos and lotions.

Choose hotels with operating windows, and save the world from the A/C.

How often do you change your sheets at home? How many towels do you go through each day? Not three. So why must you have fresh linens for every single hotel shower, nap, and night's sleep?

There are nearly 80,000 gallons of water (and five tons of waste) generated annually by each hotel room. Multiply all that by 3.3 million—the number of U.S. hotel rooms in use on an average day—and we've got a real problem. Unless you're a traveling mud wrestler, there's a simple way to make a difference: keep the same towels and sheets for three nights, and *bang*, we've saved 66% of laundry-related emissions.

InterContinental, the biggest hotel group in the world (it owns Crowne Plaza and Holiday Inn), got eco-conscious in 1990. Its initiative has grown into the International Business Leaders Forum Tourism Partnership. Founding members include Marriott, Four Seasons, Taj, and Starwood (Sheraton, W, Westin, St. Regis). Note: carbon consciousness still varies between individual properties.

It's easy to justify hotel extravagance—vacation, special occasion, road warrior tax. Unfortunately, emissions add up just the same. Good guests help hotels lower the Earth's thermostat. Here's how to help.

$$$
Marriott saved $4.5 million in energy costs with its energy-conservation program.

WRITE THE MANAGER, USE THE SUGGESTION BOX, AND ASK THAT THE HOTEL:

Allow you to keep using one set of linens

Use eco-friendly cleaning products

Use unbleached, recycled paper products

Use refillable toiletry dispensers

Turn down the A/C or heat in unoccupied rooms

Institute an energy-conservation program

Compost/donate food leftovers

Use compact fluorescent bulbs

Recycle

Don't print the bill; email it

Hotel hosedown

Each day, hotel guests use more than double the water typically used per capita at home.

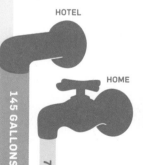

HOTEL

HOME

145 GALLONS

70 GALLONS

65 | Pack Your Go Bag

Climate change is predicted to displace millions of people by 2050. Be prepared.

COST	TIME	EFFORT	IMPACT
$30+	30 min		

Essentials

First-aid kit

Important documents (ID, map, emergency contacts, insurance policy)

Three-day supply of nonperishable food

Water: at least one gallon per person per day

Special medications; extra contact lenses or glasses

Radio and flashlight (www.redcrossstore. org sells a crank-up model with AM/FM, NOAA weather, flashlight, siren, and built-in cell-phone charger)

Waterproof matches and lighter

Tools: can opener, pocketknife, pliers (to turn off gas and water)

Toothbrush, towelettes

Cash (small bills, quarters for pay phones), credit cards

Emergency whistle

There's nothing more embarrassing than being caught unprepared for climate catastrophes.

Hurricanes, tornadoes, heat waves, flooding, drought. The latest disaster trends are just too hard to keep up with. Are white dust masks okay after Labor Day? How many emergency flashlights is too many? Does this flotation device make me look fat? Don't leave crucial questions like these to the last minute. Pack your Go Bag ahead of time and fashion a survivor look that suits you.

Think of your Go Bag as a stylish purse or weekend bag that keeps you alive for three days. (That's about how long it usually takes relief services to become operational after a disaster.) Your Go Bag is an essential part of your wardrobe and should be labeled, manageable, and accessible. Make sure everyone in the family knows where it's stowed.

Visit www.72hours.org for more information

66 | Write the Power

	COST	TIME	EFFORT	IMPACT
In the U.S., representatives receive about 200 million letters and emails per year.	$.39	5 min		

Dear Representative [THEIR NAME] :

I'm writing today to express my concerns about global warming, and to urge you to take action.

As you know, Congress has passed [NUMBER LESS THAN ONE] bills to fight global warming. This is [SYNONYM FOR "AWFUL"]. We are already the world's leading global warming polluter: the U.S. releases 7.8 billion tons of greenhouse gases a year, an increase of 16% since 1990. If we don't change our habits soon, we'll be remembered as The World's [ADJECTIVE PEJORATIVE NOUN]. (Sure, by 2009, China will surpass us in overall emissions, but that only makes the need for change more [SYNONYM FOR "URGENT"].)

I [VERB] you to support legislation in favor of public transport, increased fuel efficiency, and more research for alternative fuels. I also expect you to support recent legislation pledging the U.S. to cut its carbon emissions to 80% below 1990 levels by 2050. It's what [NAME OF FAVORITE FOUNDING FATHER] would do.

Climate change has the power to affect us all. Please help me by promising to address this issue before it's too late. If you don't take action today, our future is going to be a [ADJECTIVE] mess.

Sincerely, [YOUR NAME]

We'll need to see political action at the highest levels in order to prevent climate change. Luckily, it couldn't be easier for the average citizen to pierce the remote enclaves of national office and influence their elected representative. Just write a letter!

Visit www.congress.org for more information

67 | Pass It On

COST	TIME	EFFORT	IMPACT
$0	10 sec		

An army of millions of people fighting climate change in their own lives is good. Getting millions of others on board is much, much better.

You probably won't have to press your friends too hard to take this book. A balance of moral wisdom, frightfully dry statistics, and imaginary scenarios is already a proven winner: the three books most widely held by the world's libraries are the Bible, the United States Census, and *Mother Goose*.

Take this book and give it to a friend.

Or to a stranger. You won't just be getting more use out of the post-consumer waste you are holding in your hands. You will be on your way to helping the next generation inherit a planet to call home.

Visit www.payitforwardfoundation.org for more information

WARNING:
IF ALL ELSE FAILS

Refer to the following pages only in the event of **TOTAL CLIMATE MELTDOWN.**

These survival strategies will keep you alive on a devastated planet. They contain solutions and advice for extreme conditions. Even if it is too late to prevent climate change, it is never too late to fight for a tolerable existence.

68 | Buy a Camel

When it's time to fill up, a camel can slurp down 30 gallons of water in 10 minutes.

COST	TIME	EFFORT	IMPACT
$500	1 day		

0 gallons

of water are held in a camel's hump (or humps, if it's a Bactrian). It's fat, not water, in there.

Your dog, cat, parrot, or even boa constrictor might seem like an ideal companion for today's world. But in the not-too-distant future, the camel may become the perfect pet for the environmental challenges of the 21st century.

EASY MAINTENANCE Whether global warming brings famine or drought, your camel's good to go. Camels can survive for weeks without food and water. They can eat almost anything, including twigs and leather. Because they can raise and lower their body temperatures, camels don't sweat much, so they keep most of the water they drink in their bodies.

When desertification comes to your neighborhood, your family will be prepared. A diet of fresh camel milk will keep you alive for up to a month.

SOURCE FOR CLOTHING Bactrian camels provide a soft, warm wool from which a variety of fabrics can be made. Camel-hair coats, people. Hello?

LONG LIFE SPAN The typical camel can live as long as 50 years, far outlasting your dog or cat and rivaling most parrots and chimps.

FOOD SUPPLY Just like cows, camels can provide milk for drinking as well as for yogurt and cheese. But camel milk is more nutritious. It contains three times the vitamin C of cow's milk. It's also rich in iron, unsaturated fatty acids, and B vitamins. Some residents of the Persian Gulf even believe that it is an aphrodisiac.

RELIABLE SOURCE OF TRANSPORTATION A well-trained camel won't cover as many miles as your hybrid car, but it can travel an average of 25 miles per day at a speed of around 3 miles per hour. Without a trunk, it can still carry a load of about 600 pounds.

RECYCLABLE Camel meat provides a low-fat source of protein. It has been an Australian export since 1988. (The choicest cuts come from the hump.)

69 | Mine the Landfill

	COST	TIME	EFFORT	IMPACT
One million tons of methane gas cause 23 times the heat-trapping as the same amount of CO_2.	$0	2 hrs		

Just because it's trash doesn't mean it has to be wasted. Your municipal landfill is bursting with acres of still-useable consumer goods just waiting to be plucked out from under the steely gaze of scavenging seagulls.

There are thousands of active landfills around the world, so finding your local outlet is easy. Just head toward the bird-capped hump on the horizon, jump a fence, and scale the slopes to the top. That's where the freshest deals are deposited.

The U.S. generates more than 245 million tons of trash a year, and more than half of it gets buried. But that's only the *visible* tip of the trashberg: the energy that went into extracting, shipping, refining, manufacturing, and transporting the products that became that trash was buried along with it.

According to observers, this process could lead to an absurd scenario: it may soon be easier to reclaim mineral resources from landfills than to extract them from the earth. Already, the amount of aluminum buried in North American dumps exceeds the amount remaining in natural deposits. And there's more gold in one ton of scrap computer parts than in 17 tons of raw ore. Of course, the smartest approach is not to dump all this junk in the first place, but to find local uses or send what we can to recyclers instead. (See Skill #53.)

But to hedge your bets, consider prospecting in your local landfill.

According to the EPA, up to 5% of what currently ends up in our dumps is actually in good enough condition to be reused. That's 12.25 million tons of perfectly fine furniture, sporting equipment, appliances, and other goods buried alive every year.

70 | Beat the Heat

If greenhouse gas levels in the atmosphere reach 550 parts per million, the Earth's temperature could rise 2.9 to 5.2°F by 2100.

COST	TIME	EFFORT	IMPACT
$0	varies		

In 2003 between 35,000 and 44,000 people died in Western Europe because of an unrelenting summer heat wave; scientists have established that man-made emissions doubled the chance of such a heat wave. By 2050, more than half of Europe's summers may be as bad or worse.

BURROW

PANT

PLUCK A REED

GO NOCTURNAL

Maybe the climate skeptics are right. Instead of wasting time and money trying to prevent global warming, the best response might just be to learn to love our new tropical climate.

After all, how did we humans get to be in charge of running this planet in the first place? By being great adapters, that's how! We outwitted the Ice Age. We survived plagues and floods and earthquakes. So what if the Earth is heating up faster than at any other time in the last 10,000 years? Who's scared of a little hot weather?

These four simple actions will help keep you comfortable—even beyond 2030, when the planet's average temperature will be 1 to 2°F hotter than it is now.

BURROW We've lived in caves before—there's no shame in having to crawl back into them now. Burrowing is one of the best ways to stay cool: during summer, underground temperatures can be 30°F lower than at the surface. So grab a trowel, you've got a lot of digging to do. Your pit should be at least 30 feet deep—and the ventilation shafts are going to be a nightmare.

PANT Scientists predict that by the end of this century, global temperatures could rise by as much as 10°F. Heavy panting may become the cooling method of choice. Humans weren't made for panting (we're supposed to sweat—or live in harmony with our environment), but that's what adaptation is all about: rising to the challenge.

PLUCK A REED Nothing cools like a good, long wallow. Pluck a short reed (no longer than two feet), settle into the water, and enjoy a good chuckle at all the Chicken Littles who thought climate change meant the end of life as we know it.

GO NOCTURNAL Sometimes the best solutions are the easiest ones. Restructuring to a nighttime society is a cheap, easy fix.

Another strategy for keeping cool, developed by astronomers and supported by a grant from NASA, is to launch a "cloud" of trillions of tiny, wafer-thin satellites around the Earth. They'll cost a couple trillion dollars, but the high-flying parasols could deflect about 2% of the Sun's rays—enough to offset the warming caused by a doubling of our atmospheric CO_2.

71 | Learn to Barter

	COST	TIME	EFFORT	IMPACT
We are happy to offer you one million clay bricks in exchange for 2,000 chickens or sheep.	$0	15 min		

Ready for your daily dose of scary numbers? Oil production has already peaked in more than 50 countries, including the United States, Libya, Qatar, and Oman.

Scarier
A growing group of industry observers believes the amount of oil Saudi Arabia produces may soon go into decline.

Scarier still
Even the U.S. government believes world oil production is likely to peak "sometime between now and 2040."

One day in the not-too-distant future, things may seem…difficult. Water shortages, crop failures, displaced people, and fossil fuels too scarce or too expensive to power our world. Then what?

Well, if we've prepared a robust portfolio of alternative energy solutions, we'll be fine. On the off chance that we haven't prepared that robust portfolio, we succumb to planet-wide panic, a terrifying economic freefall, and perhaps one last, great resource war before landing with a thud in a world without energy. Once the financial markets bottom out, we get busted back to the quid pro quo economic model: two of my somethings for three of yours. It's time to test your bartering skills.

You'll have to get smart about the relative value of your remaining possessions. Otherwise, you're going to get confused and do something foolish, like barter away a family heirloom for a stick of butter. In the absence of a widely accepted medium of exchange, your wealth is directly proportionate to your natural charisma.

SURVIVAL TIPS FOR A WORLD WITHOUT CURRENCY

Spit on your possessions to make them gleam; small touches can make the difference between a good trade and a great trade.

Never make a trade solely because you heard the thing you'll be getting has magical powers. Sometimes it does, but more often you just get screwed.

Make sure to get an especially good deal before bartering away hens, draught animals, and seeds—these all have much more long-term value than you might first realize.

Be suspicious of anyone who tries to convince you to trade everything you own for an "emerging" medium of exchange. Check with your neighbors and friends before signing onto one of these "new currency" schemes. Do they also plan to adopt the new system of value? What about the people in the next village? Is the medium of exchange scarce? Is it durable?

Unfortunately, there's not much you can do if you get ripped off—there are no Better Business Bureaus or chambers of commerce in a barter economy. Your best recourse is to declare a vendetta.

Q: Who said it?
"Oil production is in decline in 33 of the 48 largest oil-producing countries."
A: Chevron

Visit www.barterco.com for more information

72 | Start a Menagerie

Climate change is predicted to endanger one million species of plants and animals by 2050.

COST	TIME	EFFORT	IMPACT
varies	40+ days		

Sixty-five million years ago our mammal ancestors burst onto the scene after massive extinctions wiped the stage clear of half the life-forms on Earth.

Now we're getting ready for our own curtain call by triggering the biggest die-off since the dinosaurs. Scientists estimate that by 2050, one-third of all plants and animals may be headed for extinction.

One way to prevent this: roll back our emissions and attempt to contain the climate changes we've already set in motion. Another way, if all that fails: travel the world collecting mating pairs of each and every species threatened by global warming.

If you're an animal lover and you've got a big backyard, you can get started today by setting up your own menagerie. A menagerie is a sort of private zoo, making it one of the only home improvements that can satisfy both your passion for conservation and your desire to make the neighbors jealous. It's not an easy venture—a menagerie of this scale will require you to spend countless hours shoveling dung—but it is a rewarding one.

Which animals should you save? The best answer, of course, is "all of them," but even with a dedicated team of roving naturalists, you'll have a hard time keeping pace with extinction rates of 50 species per day. So start with the big-name players. That means polar bears (melting Arctic sea ice), Emperor penguins (melting Antarctic sea ice), Sundarban tigers (rising sea levels), and pandas (come on, gotta have pandas).

Once you've got your Core Four, you can branch out. Consider heading to Costa Rica for a few of the cloud forests' last frogs, or Mongolia for one of the last breeding pairs of the Siberian Crane. You'll also want to think about adding a few lesser-known but still ridiculously cute animals for the kids. Try the mountain pygmy possum, the northern hairy-nosed wombat, or a furry, flower-eating, mountain-dwelling rabbit known as the pika. All of these adorable critters will probably die out relatively soon, making them perfect additions to any last-ditch collection.

73 | Harvest the Clouds

Caution: desalinating water for one million Americans with nonrenewable energy could release an extra 2.9 million tons of CO_2 per year.

COST	TIME	EFFORT	IMPACT
$400+	days		

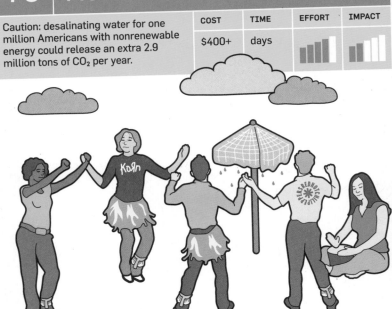

1.2 billion

people do not have access to clean water.

Human beings can live for just a few short days without water. So, when the global warming–induced water shortages get to your neighborhood, your best bet for survival is to have a few innovative H_2O solutions at the ready.

For those willing to think outside the tap, creating homemade, small-batch drinkable water is a skill worth mastering—and it's more reliable than a rain dance.

Here are three viable methods that don't require any fancy footwork.

FOG CATCHING Also known as cloud harvesting, it's simple, cheap, and effective. Large nets are stretched across hill- and mountainsides with dense fog cover. As wind blows the fog through the nets, some of the water is trapped in the mesh. Eventually, the droplets roll down the net and collect in tanks. When the nets are maintained properly, fog catchers can collect enough water daily for entire villages. Rural communities as far-flung as Chile, Yemen, and Nepal have successfully implemented this practice.

DESALINATION Some cities have huge desalination plants to remove salt from seawater. Individuals can do it on a family scale. Prices range from thousands of dollars for machines that convert hundreds of gallons per day, to much less for smaller systems that produce just a few quarts daily. (There are also small kits for campers and boaters.) Most systems use reverse osmosis, which forces seawater through a semipermeable membrane. The salt and other particles stay on one side; the desalinated water passes through to the other. Note: unless your desalination is powered by renewable energy, it will only make global warming worse.

SQUEEZING WATER OUT OF THE AIR If you combine a humidifier with a water purification system, you can make drinkable water out of thin air. A number of companies make devices that take condensed water from the air and run it through purifiers to create anywhere from 2 to 30 liters of clean water per day, depending on the humidity in the air. They aren't cheap, and most require electricity (although some are solar-powered).

During the 20th century, water consumption increased at double the rate of global population growth.

6.8 billion
gallons of water are flushed down toilets daily in the U.S.

IF ALL ELSE FAILS

74 | Build a Floating House

COST	TIME	EFFORT	IMPACT
$350k+	weeks		

If the sea level rises one meter, 100 million people in Bangladesh could be displaced.

Dutch academic Frits Schoute predicts that full-scale platforms of floating buildings will be able to accommodate living and working communities at sea, with floating cities possible by 2050.

Beachfront property is not always a good thing—especially when it's in Kansas. Global warming is already melting ice sheets and glaciers, and the resulting rise in sea levels threatens to swamp many coastal communities and islands.

Sea levels have already risen between four and eight inches in the past century, and by the end of this century, they may be three feet higher. If you want to see whether your home is at risk of being submerged by rising sea levels, check out www.flood.firetree.net.

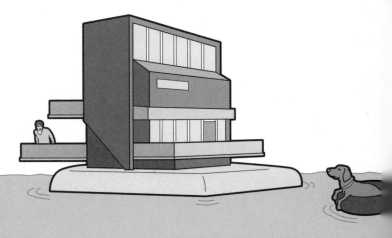

So what to do if you're at risk? Consider a floating house, like the ones created by some ingenious builders in the Netherlands.

Here's how it's done. Builders construct a corrosion-resistant platform, anchoring it deep into the seabed with pillars. The house's foundation, which sits on the platform, is a hollow concrete cube that can float. The foundation acts like the hull of a ship as water levels rise. The houses themselves are made of a lightweight, buoyant wood. To lessen the chance of floating away, the houses are moored to flexible steel posts by sliding rings that rise and fall with the water level. Electricity, water, and sewage flow through flexible pipes.

There's already a hardy community of floating houses in the small town of Maasbommel in the Netherlands. Floating houses there can withstand a rise in the water tables of up to 13 feet.

In post-Katrina New Orleans, where many homeowners already must elevate their houses by three feet, engineers are not only talking about building floating houses, but they think they can retrofit existing homes to put them on floating foundations. Engineering students from Louisiana State University have created the nonprofit Buoyant Foundation to explore alternative approaches to flood protection.

Keep in mind: the houses aren't designed to withstand hurricane force winds—they're flood-resistant, not windproof. And they're more expensive than regular homes, at least for now. On the other hand, what have you got to lose—except for your house?

634 million

people live in low-elevation coastal zones that risk flooding from rising sea levels.

Early adopters

Dutch company Oons Bouwmaatschappij has already built eight floating homes outside Amsterdam.

75 | Colonize Space

If one million climate skeptics were sent to colonize space, that would be a good start.

COST	TIME	EFFORT	IMPACT
$1,000 trillion	many years		

Potential new homes

Four local stars that are candidates for NASA's "Terrestrial Planet Finder" program:

Epsilon Indi A
About 11.8 light-years away in the constellation Indus

Epsilon Eridani
Slightly smaller and colder than the Sun, about 10.5 light-years away

Alpha Centauri B
One of the Sun's closest stellar neighbors, about 4.35 light-years away

Tau Ceti
A star long-lived enough for complex life-forms to evolve

World-famous physicist Stephen Hawking believes that if we don't colonize space, humanity could be wiped out within the next 1,000 years.

"Life on Earth is at the ever-increasing risk of being wiped out by a disaster, such as sudden global warming," Hawking says. Warming could cause the Earth's climate to become more like Venus's, he says, with sulfuric acid rain and 480°F summers.

If all else truly fails, the time may come for us to find a new home planet. Moving to another solar system—let alone next-door neighbor Mars—is by no means easy or inexpensive. The current price for a ticket to the Moon with Space Adventures, a private lunar tourism outfitter, is an astronomical $100 million per seat.

Enhance your chances of surviving the cold, dark vacuum of space by being well prepared.

BEFORE YOU GO Marry a billionaire. Escaping Earth ain't cheap. The team that took home the $10 million Ansari X Prize for building the first privately funded suborbital spacecraft had to spend more than $100 million to do it.

WHAT TO BRING FOR THE JOURNEY

A GOOD, LONG BOOK Just getting to Mars is a six-month trip each way, and reaching Proxima Centauri, our next-nearest star, would take you over four years—traveling at light speed, that is.

MULTIVITAMINS Unless your spaceship is generating its own artificial gravity, you can expect to spend the interstellar journey in a microgravity environment—during which time you'll be shedding over 1% of your bone mass each month.

POWDERED TANG Instead of bringing hundreds of thousands of gallons of freshwater into space, you'll primarily be drinking your own recycled urine, sweat, and exhaled water vapor (as well as the sweat, urine, and distilled exhalations of every other person and lab animal on board).

WHAT YOU NEED ONCE YOU GET THERE

OXYGEN TANK The Martian atmosphere has about 1/160th the oxygen of Earth's—if you somehow managed to inhale any of it, you'd end up with a lungful of CO_2. So you'll be cooped up indoors until you figure out how to create a new atmosphere from scratch.

ROLODEX To maintain your colony's genetic diversity, you should aim for a starting population of at least 1,000 healthy, breeding-age individuals.

YOUR THERAPIST Russian psychologists have found that after three months in the cramped environments of a spaceship, astronauts start to get irritable and hypersensitive.

One plan for the colonization of Mars calls for an initial massive nuclear strike directed at its polar caps. The explosion would release water vapor and CO_2—both, remember, greenhouse gases—that would then start trapping heat and other gases.

To the moon!

The U.N.'s 1979 Moon Treaty states that "the Moon and its natural resources are the common heritage of mankind." The treaty, however, has never been signed by the U.S., the UK, China, or Russia...so, if you happen to live in one of those countries, there may be a new lunar state or moon province in your country's future.

76 | Pack a Time Capsule

If one million of us buried one time capsule each, people of the future would think we were pretty weird.

COST	TIME	EFFORT	IMPACT
$0+	varies		

History books are fine for kids, but to really understand how people used to live, the best chronicle is a time capsule.

SCRATCH PAPER
"They had special paper just for throwing away? No wonder those idiots ruined my planet!"

KABADDI UNIFORM
The national sport of Bangladesh—a kind of seven-man Red Rover—will be hard up for players, as more than 15 million Bengalis will have had to flee the rising water there.

CARMEN MIRANDA–STYLE HEADPIECE
As oil reserves dwindle, shipping costs will prohibit the import of luxury items like fruit.

LACQUERED SUSHI LUNCH
Salmon populations are suffering as the temperature in their spawning grounds gets too high—no more spicy sake roll in the year 2200.

STAR TREK COLLECTIBLES
Include a few of these Worf figurines just to confuse future generations into thinking we were more advanced than we really are.

Citizens of the future will have many questions about us and the way we lived: "Why did they make my planet so hot?" "Wasn't it obviously a really bad idea to melt the ice caps?" "How could 6.5 billion people have been so stupid?" A time capsule can help answer all these questions and will give a sense of what life on Earth was like before we made it absolutely miserable. Here are some popular items to consider when assembling your global warming–themed capsule.

Decide when your time capsule should be opened, and print that date neatly on the capsule's plaque. Remember to take into account the effects of climate change—i.e., desertification, frequent flooding, and rising sea levels. A global warming capsule isn't worth much if it's underwater. The way things are going, the capsules buried during the 1939 and 1964 World's Fairs will only be accessible by sub when they're unsealed 5,000 years after they were buried.

THE STANLEY CUP
The trophy will prove that hockey teams used to skate on giant air conditioned, indoor "ice rinks."

NATURE POETRY ANTHOLOGY
Generations from now nature poetry will take the form of angry tirades directed at the merciless heat and constant flooding.

HOLIDAY SNOW GLOBE
With two to six fewer weeks of winter, you'll have to just keep dreaming of that White Christmas.

FOUR SEASONS BOXED CD SET
People in the future will know only two seasons: "regular summer" and "summer redux."

PLASTIC TOYS
In the future, free time will be spent scavenging or toiling, not playing make-believe with petroleum by-products.

IF ALL ELSE FAILS

77 | Evolve

If one million people uncoiled one strand of their DNA, the total would stretch for 631 miles.

COST	TIME	EFFORT	IMPACT
$0	many years		

The genes of at least one creature, the pitcher plant mosquito, have already shifted in response to climate change.

We've saved the most drastic option for last. In the event of catastrophic climate change, try very hard to evolve. Evolution has helped humans through plenty of tight squeezes in the past, and it's a survival strategy used by every other type of living organism. Unfortunately, it's really slow.

With a well-managed mutation program and a little luck from Mother Nature, you might be able to kick-start your personal genetic development. The results won't be perfect—you'll have to take the adaptations you get—but on an aggressive regimen of morning irradiation and afternoon mutagen shakes, you're sure to sprout something useful sooner or later.

Remember, you're hoping for any adaptation that helps you survive and pass on your mutant genes to the next generation of heat-hardy posthumans. Some changes will be more attractive than others, but you can't be too picky. Keep your eye on the ultimate goal: making sure that natural selection selects you.

GIANT EARS

Packed with blood vessels, these new mutant ears keep you cool by radiating your body heat into the surrounding air—you won't bat an eyelash when worldwide temperatures spike by 20°F. More cooling power to you if you figure out how to get them to flap.

ADAPTIVE MIGRATION

As climate patterns change, you'll start to feel an irrepressible urge to migrate toward the poles. Follow your instincts. Many species of bird—blackcaps, chiff-chaffs, and ringed plovers—have already altered their migrations to try to keep up with the shifting temperatures.

SHREWLIKE MOUTH

It currently takes 10 units of fuel energy to produce every unit of food energy. When crashing oil prices make it too expensive to grow our food on factory farms, we'll have to look to rising insect populations for dinner. Extra-long, flick-flicking tongues would be a great boon to bagging these little critters. It takes just 2,000 houseflies to make a Quarter Pounder. Tasty!

GREEN PIGMENT

You'll have to get pretty darn lucky to mutate your way into photosynthetic skin. But if you can work it, you're in the green. You'll finally have good reason to be happy about the tripling of greenhouse gases in the atmosphere expected to take place by the end of the century.

SCALY SKIN

Sure, you look like a Gila monster, but at least you're alive. By 2050, parts of the U.S. will witness 30% less rainfall. When drought strikes your suburbs, count on this tough, desiccant-proof skin to keep your innards moist.

FUR-COVERED FEET

With temperatures rising and rainfall in some places declining, it's time to face the reality that sooner or later you're going to have to cross a desert. For the best experience on the dunes, get hold of some DNA from the Sand Cat: its paws are covered with long fur to protect against the burning heat.

Afterword

Just like you, I'm learning every day how to live in a manner that reduces my impact on our climate.

On this issue, I believe we should not aim to be perfect, but to make continual progress. Reading *The Global Warming Survival Handbook* is a critical first step.

I believe we can all help tackle this problem in our own way and through our personal spheres of influence. The simple actions and skills outlined in this book, when applied on a global scale, can have a massive impact on combating global warming.

I founded Live Earth—The Concerts for a Climate in Crisis because producing, promoting, and distributing live entertainment events is what I do best, and bringing people together and triggering movements is what music does best.

My sincere hope is that July 7, 2007, the date of the Live Earth concerts, marks the beginning of a worldwide movement to combat the climate crisis. More than 100 headliners have signed on to perform for over 24 hours on all seven continents. Those artists are engaging an audience of more than two billion people across the globe through live performances and broadcasts on the Internet, television, radio, through wireless channels, and on other media. Live Earth is not only a call to action, it also provides the tools needed to answer that call with meaningful and lasting action.

Live Earth is a project of the SOS campaign, which is using all available media platforms to build a mass movement to combat the climate crisis. Global warming is the greatest environmental challenge facing humanity—not in the future and not somewhere else, but right here, right now. I urge you to tune in to Live Earth and to join the SOS movement.

Together, we can solve the climate crisis.

Kevin Wall, June 2007
FOUNDER AND WORLD PRODUCER,
LIVE EARTH AND SOS CAMPAIGN

Glossary

ALBEDO
The measure of the Sun's radiation that is reflected by the Earth's surface into space (currently about 30%)

ANTHROPOGENIC
Man-made; resulting from human influence on natural processes

CARBON CYCLE
The process by which organisms absorb and relinquish carbon from the atmosphere

CARBON DIOXIDE (CO_2)
A gas exhaled by animals and absorbed by plants; a potent greenhouse gas

CARBON DIOXIDE EQUIVALENT OR CARBON EQUIVALENT
A measure of the global warming potential (GWP) of non-CO_2 greenhouse gas emissions, expressed in terms of CO_2's GWP

CARBON SINK
Anything that emits less carbon than it absorbs from the atmosphere

CLIMATE CHANGE
A significant deviation in climate measures (such as temperature, precipitation, and wind) that lasts for decades or more

CLIMATE LAG
The delay in climate change caused by the slow effect of an event, such as the release of CO_2 into the atmosphere when the gas is stored in the ocean

DESERTIFICATION
The transformation of fertile land into a desertlike habitat, with a lowered water table, arid soil, and an increased pace of erosion

ENHANCED GREENHOUSE EFFECT
The naturally occuring greenhouse effect as exacerbated by human activity

EVAPOTRANSPIRATION
The removal of moisture from the soil by plant absorption, use, and evaporation

FEEDBACK MECHANISM
A phenomenon in which the change in one variable influences a series of events, which either weaken or strengthen climate change

FORCING MECHANISM
A process that alters the balance of the climate system, such as volcanic eruption or the enhanced greenhouse effect of CO_2

GLOBAL WARMING POTENTIAL
The measure of a gas's ability to heat the atmosphere relative to the ability of CO_2

GREENHOUSE GAS
A gas that absorbs energy in the atmosphere, such as CO_2, methane, and N_2O

IPCC (INTERGOVERNMENTAL PANEL ON CLIMATE CHANGE)
A U.N.-commissioned, international working group formed in 1988 to assess climate change and its human contributions

KYOTO PROTOCOL
An international agreement to reduce worldwide emissions of greenhouse gases, made in Kyoto, Japan, in 1997

RENEWABLE ENERGY
Power derived from resources constantly and naturally replenished on Earth

SEQUESTRATION
The storage of carbon, by trees, plants, and other materials

Resources

ALLIANCE TO SAVE ENERGY
A nonprofit policymaking coalition
www.ase.org

CENTER FOR BIOLOGICAL DIVERSITY
"Because life is good," reads the motto
of the Center, which defends endangered
species and their habitats.
www.biologicaldiversity.org

CLIMATE COUNTS
Supported by Stonyfield Farms, this
nonprofit encourages companies and consumers to fight global warming and has
an easy-to-use guide to finding clean
power in all 50 states.
www.climatecounts.org

CLIMATE PROGRESS
Frequent news and commentary on
climate change, via a project of the nonprofit Center for American Progress
www.climateprogress.org

ENVIROLINK
An enduring nexus of online resources
www.envirolink.org

ENVIRONMENTAL DEFENSE
A half-million members strong, this prag-
matic nonprofit partners with corporations
to alleviate environmental concerns.
www.environmentaldefense.org

ENVIRONMENTAL PROTECTION AGENCY
The EPA's extensive website has become an
authoritative source of ecological learning.
www.epa.gov

FRIENDS OF THE EARTH
The high-profile group has leagued together
environmentalist movements since 1969.
www.foe.org

GREEN CROSS INTERNATIONAL
Founded by Mikhail Gorbachev, GCI
works to prevent climate change, eliminate
weapons of mass destruction, and to provide safe drinking water.
www.globalgreen.org

THE GREEN GUIDE
A definitive green-consumer resource
www.thegreenguide.com

GREENPEACE
Pushing an aggressive approach to environmental change since 1971
www.greenpeace.org

GRIST
Environmental news source and blog—
as funny as it is informative
www.grist.org

LIME
A multiplatform media brand home to
a network of green-minded users
www.lime.com

NATIONAL GREEN PAGES
Products and services from almost 3,000
companies, vetted by Co-op America
www.coopamerica.org/pubs/greenpages

**NATURAL RESOURCES
DEFENSE COUNCIL**
A leading organization that uses science
and law to protect the environment
www.nrdc.org

PUBLIC INTEREST RESEARCH GROUPS
The federation of PIRGs around the U.S.
puts environmental causes near the top
of its priorities.
www.uspirg.org

SHIFTING BASELINES
A media project that brings attention to
the first climate crisis casualty, our oceans
www.shiftingbaselines.org

TREEHUGGER
The mainstream green blog hands down
sustainable style advice and news.
www.treehugger.com

UNION OF CONCERNED SCIENTISTS
The alliance of academics explains
what's behind the need for policy change.
www.ucsusa.org

WORLDCHANGING
Resources and ideas for a collaborative
eco-neutralization of human practice
www.worldchanging.org

WORLD RESOURCES INSTITUTE
Environmental think tank, celebrating
25 years of intense cogitation
www.wri.org

Global Warming Science

CLIMATE ARK
A dense and informative portal for climate
change news
www.climateark.org

THE CLIMATE INSTITUTE
The topical website is the fruit of a network of experts and alliances.
www.climate.org

**INTERGOVERNMENTAL PANEL ON
CLIMATE CHANGE**
The IPCC in Geneva is an internationally
sanctioned effort to scientifically assess
climate change.
www.ipcc.ch

**PEW CENTER ON GLOBAL
CLIMATE CHANGE**
A rich trove of global warming polling
and analysis
www.pewclimate.org

**UNIVERSITY CORPORATION FOR
ATMOSPHERIC RESEARCH**
The heavy-hitting research institute in
Boulder, Colorado, is a reliable monitor
of climate change.
www.ucar.edu/research/climate

Business

1% FOR THE PLANET
Leagued together by two conservationist
businessmen, members donate 1%
of sales to environmentalist programs.
www.onepercentfortheplanet.org

GREENBIZ & CLIMATEBIZ
Help companies become environmentally
responsible without compromising profits
greenbiz.com
climatebiz.com

SOCIAL INVESTMENT FORUM
Staunch advocate and resource for socially
responsible investing
www.socialinvest.org

Food, Shelter & Travel

BUILDING GREEN
Clearinghouse of green building information
www.buildinggreen.com

EARTHSCREEN
Savings and rewards for earth-friendly
shopping
www.earthscreen.com

ENERGY STAR
Educate yourself about the products and home building methods that meet EPA and U.S. DOE energy guidelines
www.energystar.gov

GREENHOMEBUILDING
Learn about sustainable building
www.greenhomebuilding.com

HOTEL ECO-LABELS
Among the organizations that certify hotels' ecofriendliness are Green Globe, Green Seal, EcoClub, and Project Planet.
www.greenglobe.org
www.greenseal.org
www.ecoclub.com
www.projectplanet.biz

LOCALHARVEST
A "buy local" website with a huge directory of farmer's markets
www.localharvest.org

ORGANIC CONSUMERS ASSOCIATION
News for America's estimated 50 million organic customers
www.organicconsumers.org

SUSTAINABLE TRAVEL INTERNATIONAL
Promotes eco-friendly practices among travelers and travel-related companies
www.sustainabletravelinternational.org

U.S. GREEN BUILDING COUNCIL
The foremost coalition of building professionals working to create environmentally responsible places to live and work
www.usgbc.org

Transportation

NATIONAL CENTER FOR BICYCLING & WALKING
Policy planning and support for like-minded transportation organizations
www.bikewalk.org

GREENERCARS
Top green car choices, from the American Council for an Energy-Efficient Economy
www.greenercars.com

ERIDESHARE
The No. 1 online bulletin for rideshares and slugging in North America
www.erideshare.com

FUELECONOMY.GOV
The EPA's one-stop shop for fuel ratings
www.fueleconomy.gov

THUNDERHEAD ALLIANCE
A national coalition of state and local cycling advocacy groups
www.thunderheadalliance.org

Books

PLAN B 2.0: RESCUING A PLANET UNDER STRESS AND A CIVILIZATION IN TROUBLE
Lester R. Brown (Norton, 2006)
A detailed prescription for the Earth's environmental challenges, by the renowned founder of the Earth Policy Institute

AN INCONVENIENT TRUTH
Al Gore (Rodale, 2006)
Vice President Al Gore's slideshow-turned-movie-turned-book presents the facts about climate change with a heartfelt narrative and intuitive graphics.

WORLDCHANGING: A USER'S GUIDE FOR THE 21ST CENTURY
Alex Steffen, ed. (Abrams, 2006)
The compendium of the present-day green movement is biblical in scope and size.

THE GREEN HOUSE: NEW DIRECTIONS IN SUSTAINABLE ARCHITECTURE
(Princeton Architectural Press, 2005)
A survey of the most beautiful and most compelling sustainable homes from around the world

FIELD NOTES FROM A CATASTROPHE: MAN, NATURE, AND CLIMATE CHANGE
Elizabeth Kolbert (Bloomsbury, 2006)
A clear, concise survey of the already apparent effects and the coming challenges of climate change

NATURAL CAPITALISM: CREATING THE NEXT INDUSTRIAL REVOLUTION
Paul Hawken, Amory Lovins, and L. Hunter Lovins (Back Bay, 2000)
Expert perspective on transforming the business world to address ecological crises

Behind This Book

Bringing together more than 150 artists on seven continents, the Live Earth concerts use the global reach of music to sound a call to action.
www.liveearth.org

Save Our Selves—The Campaign for a Climate in Crisis will trigger a global movement to combat our climate crisis, beginning with the Live Earth concerts.
www.saveourselves.com

ADVENTURE ECOLOGY

Adventure Ecology is a youth movement that provides an inspirational platform for change, community, and action on global warming.
www.adventureecology.com

Brand Neutral

Brand Neutral teaches companies environmental and energy strategies to cut costs and reach consumers.
www.brandneutral.com

THE °CLIMATE GROUP

In 2007, The Climate Group, an independent, nonprofit organization, launches the "We're in This Together" campaign, a partnership of companies who aim to make it easier for their customers to take action on climate change.
www.theclimategroup.org
www.together.com

Acknowledgments

Live Earth would like to thank everyone at Live Earth and Adventure Ecology.

This book would not have been possible without Leigh Haber at Rodale and the dedicated team at Melcher Media: Charles Melcher, publisher; Bonnie Eldon, associate publisher; Duncan Bock, editor in chief; David E. Brown, editor; Lindsey Stanberry, assistant editor; Kurt Andrews, production director; and Jessi Rymill and Carl Williamson, designers at Circle & Square.

Melcher Media would like to thank Bronwyn Barnes, Adam M. Bright, Morgan Clendaniel, Daniel Del Valle, Max Dickstein, Alissa Faden, Tom Feegel, Holly Gressley, Rachel Griffin, Kit Hawkins, Charles Komanoff, Lisa Maione, Mark Miller, Betty Morin, Andrea Nakayama, Lauren Nathan, Kirthana Ramisetti, John Rego, Lia Ronnen, Holly Rothman, Kristina Schake, Ellen Sitkin, Lily Sobhani, Josh Stempel, Alex Tart, Shoshana Thaler, Marti Trgovich, Betty Wong, and Megan Worman.

DESIGN & INFOGRAPHICS BY Circle & Square

ILLUSTRATIONS BY William van Roden

PRODUCED BY

 MELCHER MEDIA

124 West 13th Street
New York, NY 10011
www.melcher.com

Melcher Media strives to use environmentally responsible suppliers and materials whenever possible in the production of its books. For this book, that includes the use of vegetable-based inks printed on Revive 100 Offset, a 100% post-consumer waste paper stock with Forest Stewardship Council certification. The cover of this book is laminated with a water-based, biodegradable coating.

With a product to suit every print application, environmental requirement, and budget constraint, Revive is the UK's leading recycled paper range. Available exclusively from the Robert Horne Group, www.roberthorne.co.uk.

This is a carbon-neutral book: with the help of NativeEnergy, 100% of the carbon emissions from the energy used to make this book are being offset by investments in new renewable-energy projects. For more information on how this book was offset and to address your own carbon footprint, visit www.NativeEnergy.com.